Lecture Notes in Mathematics

A collection of informal reports and seminars
Edited by A. Dold, Heidelberg and B. Eckmann, Zürich

Series: Forschungsinstitut für Mathematik, ETH, Zürich · Adviser: K. Chandrasekharan

33

György I. Targonski
Fordham University, New York

Seminar on Functional Operators and Equations

Forschungsinstitut für Mathematik, ETH, Zürich
October 1965 – July 1966

1967

Springer-Verlag · Berlin · Heidelberg · New York

Master copy prepared by Mrs. Jacqueline Ellis

All rights, especially that of translation into foreign languages, reserved. It is also forbidden to reproduce this book, either whole or in part, by photomechanical means (photostat, microfilm and/or microcard) or by other procedure without written permission from Springer Verlag. © by Springer-Verlag Berlin · Heidelberg 1967
Library of Congress Catalog Card Number 67-24323 Printed in Germany. Title No. 7353.

TABLE OF CONTENTS

		Page
Introduction and Summary		1
1.	Multiplication Theorems and Square Theorems; Bourlet Operators	5
2.	Substitutions Operators. The Schröder Equation	21
3.	Continuous Iteration. Commuting Substitution Operators	31
4.	Relations Between Bourlet Operators of the Specific Form	45
5.	Superposition of Substitution Operators	56
6.	Integral Operators: Remarks and Definitions	61
7.	The Theorems of Weyl and von Neumann on Hermitean Carleman Operators	70
8.	Recent Results on Carleman Operators	75
9.	Strong Carleman Operators	86
10.	Convergence Theorems	92
11.	Transformation of Strong Kernels	99
12.	Operators of Locally Bounded Range	102
13.	Concluding Remarks. Open Questions	104
List of References		108

INTRODUCTION AND SUMMARY

These notes contain the material - with some results added - of a seminar I had the opportunity to give at the Forschungsinstitut für Mathematik der Eidgenössischen Technischen Hochschule, Zürich, in the academic year 1965-66, entitled: Funktionaloperatoren und-Gleichungen.

My sincere thanks are due to B. Eckmann for hospitality at the Forschungsinstitut during the year, and to the participants of the seminars, especially to H. Diamond, J. Lambek and F. W. Lawvere for many stimulating discussions.

The title of the seminar is taken from S. Pincherle's Encyclopedy article [2] to emphasize that - at least in the first half of the material covered - ideas stemming from the "naive" period of operator theory are followed up in the spirit of newer developments.

This first part starts with C. Bourlet's idea of linear operators with a multiplication theorem

$$\Omega(uv) = F[u, v, \Omega u, \Omega v]$$

where u and v are elements of a commutative algebra, in particular a function algebra with a point-wise multiplication. It turns out that - at least in algebras with unit element and without divisors of zero - such operators essentially belong to one of three types: they are derivations, or substitutions, or multipliers. (Sec. 1). Looking at the relations between these "Bourlet operators", i.e. linear operators with a multiplication theorem, it turns out, among other things, that derivations are, crudely speaking, logarithms of substitutions (Sec. 4).

The eigenvalue problem for Bourlet operators leads - in the case of substitutions to the Schröder Equation. (Sec. 2). This points to an analogous treatment of first-order differential operators

$$f(x) \to \omega(x) \frac{d}{dx} f(x)$$

and substitution operators

$$f(x) \to f[\omega(x)].$$

In particular, the question of representability of arbitrary functions in terms of the solution of a given Schröder equation is analogous to the Weyl-Ikebe representation problem for differential operators. The situation is demonstrated using an algebra of polynomials. (Sec. 2).

Families of commuting substitution operators make use of the solutions of the Schröder equation in a different direction. A generalization of the Chebyshev polynomials arises here in a natural fashion. (Sec. 3).

Superposition of substitution operators leads to integral operators (Sec. 5) and this idea forms a connecting piece to the second semester material of the seminar: Carleman integral operators.

Integral operators - more precisely, the integral representation of linear operators - are first discussed in a general setting. Then Carleman integral operators on the function space L^2 are chosen as the subject of subsequent discussion. (Sec. 6).

As a basis for generalization, the theorems of Weyl and von Neumann on hermitean Carleman operators are discussed (Sec. 7). Generalizations to normal and in some cases non-normal operators are then given, the crucial question always being (to borrow a phrase coined by J. B. Miller [28] for a related situation): To what extent is the Carleman property "contagious"?, i.e. whether Ω^*, $U\Omega U^*$, $A\Omega$, ΩB is a Carleman operator if Ω is one. (Sec. 8). These problems, originating with Weyl, Carleman, von Neumann and other mathematicians, recently taken up by M. Schreiber and, in a different direction, by B. Misra, D. Speiser and myself, seem to have interested mainly physicists for quite a time. In fact, the investigation which led to the results described here was suggested by J. M. Jauch in connection with scattering theory.

"Strong" Carlemann operators I call those which preserve their Carleman property under arbitrary unitary transformations; the algebra of such operators is isomorphic to the algebra of the corresponding kernels. It is shown how the basic operations of the operator algebra appear in the algebra of kernels and a number of related results given. In particular, it is proven that a bounded, strong Carleman operator is a Hilbert-Schmidt operator (in the generalized sense). (Sec. 9).

Somewhat surprisingly, the theory of Carleman operators turns out to be related to the question of convergence almost everywhere of the series $\sum |c_n \phi_n(x)|^2$ where $\{\phi_n\}$ is an orthonormal system. Theorems in this direction are derived. (Sec. 10).

Returning to the question of unitary transformation of strong kernels, we find that the problem becomes interesting in its own right and some results about the space $\underline{L}^2$ of functions of two variables result. (Sec. 11).

The material continues with a lemma about Carleman operators somewhat aside from the previous line of thought (Sec. 12), and it concludes with a list of unsolved problems arising from several sections, as well as comments on some related topics. The present text is a permanent version of the hectographed notes distributed to the participants of the seminar.

1. MULTIPLICATION THEOREMS AND SQUARE THEOREMS; BOURLET OPERATORS

A "constructive theory of linear operators" should deal with approximation and expansion of linear operators in terms of "simple" linear operators, in an analogy to the constructive theory of functions. We know that polynomials are simple in a fundamental sense, trigonometric, Bessel, etc. functions simple at least in a historic sense. But when is a linear operator on an abstract space "simple"? We shall seek the answer in demanding that beside linearity, Ω should have some simple multiplicative property, i.e. we must be able to say something about $\Omega(uv)$; this indicates that we cannot formulate our criterion in terms of a space; we must have an algebra.

Let $u, v, \ldots$ then be an algebra, the elements of which are (real or complex valued) functions of one (real or complex) variable. We do not yet impose any norm on our algebra. To define "simplicity", we follow a suggestion of C. Bourlet, made in 1897 [1], and demand that $\Omega(uv)$ should be expressed in terms of u, v, Ωu, Ωv only. In a more modern language, $\Omega(uv)$ should have a "multiplication theorem".

(1.1) $\qquad (uv) = F[u, v, \Omega u, \Omega v].$

(1.1) has to be understood in the following sense. There exists a unique function F of four variables such that for every pair u, v in our algebra and for any fixed value of the variable in the domain of definition (1.1) should be a numerical equality. Later, we shall show that F must be a very special function and that there exist essentially only three multiplication theorems:

(1.2a) $\quad\quad\quad\quad \Omega(uv) = A(\Omega u)(\Omega v)$

(1.2b) $\quad\quad\quad\quad \Omega(uv) = \frac{1}{2}[u\Omega v + v\Omega u]$

(1.2c) $\quad\quad\quad\quad \Omega(uv) = u\Omega v + v\Omega u.$

Operators satisfying a multiplication formula of one of these types ("Bourlet operators") have, e.g. on a polynomial algebra, essentially the explicit form

(1.3a) $\quad\quad\quad\quad (\Omega f)(x) = Af[\frac{1}{A}\omega(x)] - Cf(x)$

(1.3b) $\quad\quad\quad\quad (\Omega f)(x) = \Omega(x)f(x)$

(1.3c) $\quad\quad\quad\quad (\Omega f)(x) = \omega(x)\frac{d}{dx}f(x) + \lambda f(x),$

where $\omega(x)$ is an element of our algebra. This result was also anticipated by Bourlet (cf. also [2], p. 431).

The "naive" period of operator theory, i.e. the late nineteenth and early twentieth centuries, produced several such ideas, which later were half-forgotten essentially because they can be formulated rigorously only in the framework of a Banach algebra theory.

Operator (1.3b) is simpler than the other two, (1.3c) has a very well developed theory, but what about (1.3a). It is interesting for at least four different reasons.

1. In our approach it appears on equal footing with the first-order differential operator.
2. For the case $A = 1$, it is a linear ring endomorphism of our algebra. (We shall even show that it is the only possible linear ring endomor-

phism for polynomials, power series etc. algebras.)

3. It made its appearance in mathematical physics as early as 1931 (in the work of B. O. Koopman) [10] through the operation

(1.4) $$\Omega_t f(P_o) = f(S_t P_o) = f(P_t)$$

where P_o is a point in a (finite or countably infinite dimensional) phase space describing the state of a physical system as time $t = 0$, S_t an operator carrying the system into its state as time t, f a scalar function ("measurement") of the state of the system at a given time. Obviously, Ω_t in (1.4) has the property (1.2a). The theory, however, was not pursued further since products like $f_1(P)f_2(P)$ appeared to have no physical meaning – at least, in quantum theory.*

4. The most important reason for the study of operators of the form (1.3a) is that they play a fundamental role in the theory of functional equations. For example, the Kordylewski-Kuczma equation [3]

$$\phi[\omega(x)] + g(x)\phi(x) = F(x)$$

[ϕ unknown] contains in the first term an operator in this form. In addition, this type of functional equation is, in a sense, the most fundamental one, the one to which many others can be reduced. We shall say more about this when we come to the Schröder equation.

A study of the operator (1.3a) is, in any case, as important for the understanding of Functional Equations as the study of differential operators (1.3c) is for the understanding of Differential Equations. The analogous

---

*I am indebted to L. Radicati for a discussion clearing this point.

properties of (1.3a) and (1.3c) - which led to the original observation of Bourlet - makes the proposed study both important and relatively easy.

An additional reason for the investigation of "Bourlet operators" is their relationship to smoothing operators as recently studied by G.-C. Rota [4]. Such operators - characterized earlier by "smoothing properties" as, for example, the decrease in total variation after application of the operator - have been recently characterized algebraically through the Reynolds identity,

(1.5) $\qquad R(uv) = RuRv + R[(u-Ru)(v-Rv)]$.

While (1.5) is not a multiplication theorem in the sense of (1.1), a simple investigation reveals the fact that smoothing operators are essentially the inverses of Bourlet operators.

The eigenvalue and eigenfunction problem for operators (1.3a) leads to the equation

(1.6) $\qquad (\Omega\phi)(x) = \phi[\omega(x)] = \lambda\phi(x)$, $\qquad \phi$ unknown,

i.e. Schröder's Equation, which has received very much attention from the original paper by Schröder in 1871 [5] to the study of Szekeres [14] in 1959 and the even newer results by Kuczma [15].

We shall group most of our questions regarding the operators of type (1.3a) in the following framework, and we shall refer to the following questions as three "principal questions".

I. Is, in a given function algebra, <u>every</u> linear "ring endomorphism"

$$\Omega(u+v) = \Omega u + \Omega v, \qquad \Omega(uv) = \Omega u \Omega v$$

a substitution operator of type (1.3a)?

II. Does there exist, in a given function algebra, a solution for the Schröder Equation

$$\phi[\omega(x)] = \lambda\phi(x)$$

where ω belongs to the algebra? If not, for what sub-sets in the algebra does a solution exist?

III. Provided a solution exists, can one represent an arbitrary element of the algebra in the form

(1.7) $$f(x) = \sum_{n=0}^{\infty} A_n \phi(x)^n ?$$

Here we have used the fact that, together with $\phi(x)$, $\phi(x)^n$, (n = 0,2,3,4) is also a solution - a fact which facilitates the treatment. (Series of type (1.1), Bürmann series, are treated by Sansone-Gerretsen [5].)

We shall generalize question III. by introducing a "ξ-algebra", a function algebra which contains $|\phi(x)|^\xi$, $\xi > 0$, if it contains ϕ. Using this notion, (1.7) is generalized to

(1.8) $$f(x) = \sum_{n=0}^{\infty} A_n \phi(x)^n + \int_0^{\infty} A(\nu) |\phi(x)|^\nu d\nu.$$

This representation is analogous to the Weyl-Ikebe representation in terms of eigenfunctions of differential operators [7].

We shall also investigate one-parameter semi-groups of Bourlet operators using continuous iteration, an analogue of fractional differentiation. Then we shall investigate the representation of general linear operators as the sum (or integral) of Bourlet operators in the form

$$(1.9) \qquad A = \sum_{n=0}^{\infty} a_n \Omega^n + \int_0^{\infty} d\nu a(\nu) \Omega^\nu,$$

where Ω is a Bourlet operator, i.e. one of the three types from among (1.3a), (1.3b) (1.3c), and, of course, the second term in (1.9) has to be carefully defined.

Representations of the form (1.9) will lead us in a natural way to integral operators or, more precisely, the integral form of operators, since we shall regard the integral operator as a "standardized way" of writing a general linear operator. Wider and wider classes of operators can be expressed in this "standardized" form

$$(1.10) \qquad (Af)(x) = \int_a^b a(x,y)f(y)dy$$

as we gradually relax conditions imposed on the kernel $a(x,y)$. In particular, we shall discuss in detail the Carleman operators and the narrower class of Hilbert-Schmidt operators.

Throughout, we defined multiplication in our function algebras as ordinary (point-wise) multiplication. We now consider operators with a multiplication theorem. As we said previously, we shall refer to them as Bourlet operators.

Let Ψ be a commutative algebra with unit element and without divisors of zero (See [8], pp. 19-22 and [9], p. 2.) We note that Ψ is not necessarily a Banach algebra, i.e. not necessarily complete in some norm. On this algebra we now define Bourlet operators Ω, satisfying

[1.1], and mapping Ψ into itself. F is understood to be a (rational or transcendental) entire function in the four variables, i.e.

$$(1.11) \qquad F = \sum_{k,\ell,m,n=0}^{\infty} A_{k\ell mn} u^k v^\ell (\Omega u)^m (\Omega v)^n$$

where the r·h·s is convergent in the norm.

Let us add three remarks.

First. Ψ is not complete in the norm, so an apparent contradiction arises from the fact that $uv \in \Psi$. Thus $\Omega(uv) \in \Psi$, while the r·h·s as the sum of an infinite series does, apparently, not always belong to Ψ. We shall however see that F is a special polynomial, thus the problem does not arise.

Second. We could have considered, instead of (1.1), a "square theorem",

$$(1.12) \qquad \Omega(u^2) = G(u, \Omega u).$$

The knowledge of this is equivalent with the knowledge of (1.1). Obviously, the choice of $v = u$ in (1.1) leads to (1.12). Conversely, writing v for u in the "square theorem" (1.12), and subtracting we find, after introducing

$$u + v = U, \quad \text{and} \quad u - v = V$$

$$\Omega(UV) = G\left(\frac{U+V}{2}, \frac{\Omega U + \Omega V}{2}\right) - G\left(\frac{U-V}{2}, \frac{\Omega U - \Omega V}{2}\right),$$

i.e. a multiplication theorem.

Third. What is the significance of having a multiplication theorem? We set out from the very beginning by considering "constructivity". Let be $\phi \in \Psi$ some element such that the elements of Ψ can be expanded, or in some way approximated in polynomials in ϕ. Then, in order to "know" Ω, it is sufficient to know the effect of Ω on the powers of ϕ. But by (1.1)

(1.13) $$\Omega(\phi^n) = \Omega(\phi^{n-1} \cdot \phi) = F[\phi^{n-1}, \phi, \Omega(\phi^{n-1}), \Omega\phi].$$

Thus we have a recursion for the determination of $\Omega(\phi^n)$.

F is of very special from. From (1.1), because of the homogeneity of Ω, in (1.11) only the quadratic terms $k + \ell + m + n = 2$ differ from zero. Moreover, since u and v play a symmetric role (Ψ being a commutative algebra), the corresponding coefficients are equal and

(1.14) $$F = A_{2000}(u^2 + v^2) + 2A_{1100}uv + A_{0020}[(\Omega u)^2 + (\Omega v)^2]$$
$$+ 2A_{0011}[(\Omega u)(\Omega v)] + 2A_{1010}[u\Omega u + v\Omega v]$$
$$+ 2A_{1001}[u\Omega v + v\Omega u] = \Omega(uv).$$

First putting $u = 0$, and then putting $v = 0$ in (1.14), we find

(1.15) $$A_{2000}u^2 + A_{0020}(\Omega u)^2 + 2A_{1010}u\Omega u = 0,$$

and correspondingly for v. Thus (1.14) becomes

(1.16) $$\Omega(uv) = A(\Omega u)(\Omega v) + B[u\Omega v + v\Omega u] + Cuv.$$

Now let $A \neq 0$. Then the system

(1.17) $$1 - A(\lambda + \mu) = 2B \quad , \quad A\lambda\mu = C$$

has a solution for all B,C. We substitute (1.17) in (1.16) and choose $u = v = e$ (e being the unit element of ψ and we write $\Omega e = e_o$) and obtain

(1.18) $$Ae_o^2 - A(\lambda+\mu)e_o + A\lambda\mu e_o = 0, \text{ or}$$

$$(e_o - \lambda e)(e_o - \mu e) = 0,$$

since $A \neq 0$. Since Ψ has no divisors of zero, all operators Ω satisfying the multiplication theorem (1.16) with given A, B, C, therefore given λ, μ belong in two classes; either $\Omega e = e_o = \lambda e$ or $\Omega e = e_o = \mu e$. Let us assume that our operator satisfies $e_o = \lambda e$, and choose $v = e$ in (1.16). Then

(1.19) $$[1 + A(\mu-\lambda)]\Omega u = \lambda[1 + A(\mu-\lambda)]u.$$

For $[1 + A(\mu-\lambda)] \neq 0$ this means that $\Omega u = \lambda u$, and for $[1 + A(\mu-\lambda)] = 0$, $\lambda = \mu + \frac{1}{A}$, i.e. from (1.16) and (1.17)

(1.20) $$(\Omega-\mu I)(uv) = A[(\Omega-\mu I)u(\Omega-\mu I)v].$$

Introducing the operator

(1.21) $$\Omega - \mu I = \Omega_o,$$

(1.20) becomes

(1.22) $$\Omega_o(uv) = A\Omega_o u \Omega_o v.$$

To summarize, for $A \neq 0$ in (1.10), we have either $\Omega u = \lambda u$ or (1.22).

For $A = 0$, (1.16) yields with $u = v = e$ (now the absence of divisors of zero in Ψ is not used),

(1.23) $(1 - 2B)e_0 = Ce.$

For $B = \frac{1}{2}$, this means $C = 0$, i.e.

(1.24) $\Omega(uv) = \frac{1}{2}[u\Omega v + v\Omega u].$

For $B \neq \frac{1}{2}$, $e_0 = \frac{C}{1-2B} e$ follows, i.e. from (1.16), putting $v = e$

(1.25) $(1-B)\Omega u = \frac{C(1-B)}{1 - 2B} u.$

Again, for $B \neq 1$, $\Omega u = \frac{C}{1-2B} u$. For $B = 1$, we have

(1.26) $\Omega(uv) = u\Omega v + v\Omega u + Cuv.$

Introducing the operator

(1.27) $\Omega_1 = \Omega + CI,$

(1.26) becomes

(1.28) $\Omega_1(uv) = u\Omega_1 v + v\Omega_1 u.$

To summarize: for $A = 0$, Ω is either a multiple of the identity operator, or one of the multiplication theorems (1.24), or (1.28) is satisfied for an operator resulting from Ω by subtracting some multiple of the identity. We can drop the case where Ω is a multiple of the identity since then Ω satisfies (1.24), as we shall see at once.

We shall henceforth distinguish the three possible "types" of Bourlet operators as follows:

    (1.22)   Type D

    (1.24)   Type M

    (1.28)   Type H.

Since we are practicing a "constructive" approach, we now ask the question: What is the "explicit" form of operators of type H, M, or D? The question can be at once answered in full generality in the case of type M; the two others we shall investigate in special algebras.

Write $v = e$ in (1.24). Then

(1.29) $\qquad \Omega u = e_o u$

i.e., an operator of type M is always a multiplication by a fixed element of Ψ namely by the image of the unit element.

Trying a similar approach on type D, we merely find $e_o = 0$, and no other information on the operator. We summarize our findings in

Theorem 1.1. *Let Ψ be a commutative algebra, and consider linear operators Ω mapping Ψ into itself and satisfying a "multiplication theorem"*

$$\Omega(uv) = F[u,v,\Omega u,\Omega v]$$

where F is an entire function in four variables. Then F is necessarily of the form

$$F = A(\Omega u)(\Omega v) + B[u\Omega v + v\Omega u] + Cuv.$$

If Ψ has a unit element and no divisors of zero, every such operator is either a multiple of the identity or belongs to one of the following three types:

H: $\qquad \Omega_o(uv) = A\Omega_o u \Omega_o v \qquad \Omega_o = \Omega - \mu I$

M: $\qquad \Omega(uv) = \frac{1}{2}[v\Omega u + u\Omega v]$

D: $\qquad \Omega_1(uv) = v\Omega_1 u + u\Omega_1 v \qquad \Omega_1 = \Omega + CI.$

An operator of type M is always equivalent to a multiplication by a fixed element, and an operator of type D always annihilates the unit element.

Let us now return to formulae (1.3a), (1.3b) and (1.3c). We derive these on the algebra P of all polynomials of a complex variable. There are no divisors of zero and $e = p(z) \equiv 1$ is the unit element. Addition and multiplication shall be the ordinary "point-wise" operations.

Let Ω be a Bourlet operator mapping P into itself. Theorem 1.1 is valid in P and we are now looking for the excplicit form of the operators of type D and H. For type M we already know

(1.30) $\qquad (\Omega p)(z) = g(z)p(z)$, for all $p \in P$

where $g(z)$ denotes a fixed polynomial independent of p.

For the case of type D, we know from Theorem 1.1 that, for a suitable C, $\Omega_1 = \Omega + CI$ satisfies the multiplication theorem $\Omega_1(uv) = v\Omega_1 u + u\Omega_1 v$. Putting here $u = v = e$, we find, as already seen, $\Omega_1 e = 0$. Moreover, we define

(1.31) $\qquad \Omega_1 z = \omega_1(z)$.

Now we prove

(1.32) $\qquad \Omega_1(z^k) = k\omega_1(z)z^{k-1} = \omega_1(z)\dfrac{d}{dz}z^k$

by induction. Indeed, from the multiplication theorem, using (1.31), and assuming the correctness of (1.31) for k,

$$\Omega_1(z^{k+1}) = \Omega_1(z \cdot z^k) = z\Omega_1(z^k) + z^k \Omega_1(z)$$
$$= zk\omega_1(z)z^{k-1} + z_k\omega_1(z) = (k+1)\omega_1(z)z^k,$$

and because of $\Omega_1 e = 0$, the formula is correct for $k = 0$. It follows that for every polynomial $p(z) \in P$, $(\Omega_1 p)(z) = \omega_1(z) \frac{d}{dz} p(z)$ and ultimately

(1.33) $\qquad (\Omega p)(z) = (\Omega_1 - CI)p(z) = \omega_1(x) \frac{d}{dz} p(z) - Cp(z).$

In the algebra P, therefore, every operator of type D is a first-order differential operator of the form given in (1.33).

In the case of type H, we know that for a suitable μ, $\Omega_o = \Omega - \mu I$ satisfies the multiplication theorem $\Omega_o(uv) = A\Omega_o(u)\Omega_o(v)$. For arbitrary $v = e$ and arbitrary u, this yields $\Omega_o u = A\Omega_o u \Omega_o e$. Thus, unless Ω_o is the zero operator,

(1.34) $\qquad \Omega_o e = \frac{1}{A} e.$

We now define

(1.35) $\qquad \Omega_o z = \omega_o(z) \in P.$

We will prove by induction,

(1.36) $\qquad \Omega_o(z^k) = A^{k-1} \omega_o(z)^k = \frac{1}{A}[A\omega_o(z)]^k, \quad k \geq 1.$

Indeed, assuming (1.36) to be correct for some $r \geq 1$, the multiplication theorem yields, using (1.35)

$$\Omega_o(z^{k+1}) = \Omega_o(z \cdot z^k) = A\Omega_o(z)\Omega_o(z^k)$$
$$= A\omega_o(z)A^{k-1}\omega_o(z)^k = A^k \omega_o(z)^{k+1}.$$

From (1.34) and (1.36) we find for every polynomial

$$(\Omega_o p)(z) = \frac{1}{A} p[A\omega_o(z)],$$

i.e., because of $\Omega_o = \Omega - \mu I$ and

(1.37) $\qquad (\Omega p)(z) = \frac{1}{A} p[A\omega_\sigma(z)] + \mu p(z).$

For $A = 1$ and $\mu = 0$, Ω is a substitution operator in the strict sense. In physics such operators are often referred to as generalized scale transformations. (Scale transformations in the strict sense are linear transformations of the independent variable, i.e. in our case $\omega_o(z) = \alpha z + \beta$.) The interesting point here is exactly that non-linear transformations on the variable space induce linear transformations on the algebra of functions defined on this space, moreover, linear transformations obeying a simple multiplication theorem, namely, $\Omega(uv) = \Omega u \Omega v$. (cf. also [10]).

Let us introduce, at this point, the following terminology. Differential operators of the form (1.33), generalized substitutions operators of the form (1.37), and multipliers of the form (1.30) will be called, on any function algebra with point-wise multiplication, Bourlet operators in the specific form (of type D, or H, or M, respectively). It is clear that every Bourlet operator in the specific form is, a fortiori, a Bourlet operator.

The question is obviously to be asked, whether a Bourlet operator on a commutative algebra is necessarily of the specific form. This question corresponds exactly to the first of the three principal questions. Let us call it the "reversal question". There, the question is formulated for type H only, with good reason. For type M we already know from Theorem 1.1 that the answer to the reversal question is in the affirmative. An operator of type M is always a multiplier. For type D, the question arises, what happens in an algebra containing non-differentiable elements. We are concerned, however, with differential operators only as long as some analogy with operators of type H is investigated.

For the class H, the difficulty of the reversal question arises in two steps. On a function algebra, there certainly exist H-type operators in the specific form, i.e. substitutions in the generalized form (1.37). The question is, whether there exist other operators of type H which cannot be written in the specific form. As seen, for the polynomial algebra no such operators exist. The next two algebras to be investigated would be the algebra C of continuous functions on a compact set. (Perhaps, one should start with the Stone-Weierstrass theorem.)

Using the "sup" norm, and the algebra of entire functions of one complex variable

$$f(z) = \sum_{n=0}^{\infty} A_n z^n,$$

using the norm (cf. [9], pp. 317-318)

$$\|f\| = \sum_{n=0}^{\infty} \alpha_n |A_n|.$$

Here $\{\alpha_n\}$ is a sequence of positive numbers with $\alpha_0 = 1$, $\alpha_{m+n} \leq \alpha_m \alpha_n$ and $\lim_{n \to \infty} (\alpha_n)^{1/n} = 0$. This algebra, however, is not complete in the norm. The question is settled, however, in the case of P.

We summarize our findings in the following lemma.

<u>Lemma 1.1.</u> <u>On the algebra of all complex polynomials of one complex variable, all Bourlet operators are of the specific form</u>

$$(\Omega p)(z) = \frac{1}{A} p[A\omega_o(t)] - Cp(z) \qquad \text{Type H}$$

or

$$(\Omega p)(z) = \omega(z) \frac{d}{dz} p(z) + \mu p(z) \qquad \text{Type D}$$

or

$$(\Omega p)(z) = \omega(z) p(z) \qquad \text{Type M.}$$

2. SUBSTITUTION OPERATORS. THE SCHRÖDER EQUATION

Let us now consider operators of type H of the specific form, in the simple special case $A = 1$, $C = 0$,

(2.1) $\quad\quad Hf(x) = f[\omega(x)]$,

i.e. substitution operators. The eigenvalue equation of this operator is the Schröder equation

(2.2) $\quad\quad f[\omega(x)] = \lambda f(x)$.

This equation was introduced by E. Schröder in 1871 [5] in a more general form and has been the subject of many investigations. From among the most recent ones we mention the work of Szekeres [14] and the very complete paper by Kuczma [15]. Schröder's equation was introduced because of its connection with the problem of continuous iteration, a problem to which we will turn in the next section. In one approach, however, the Schröder equation appears as an eigenvalue equation. The two approaches are, of course, related.

We first notice that the spectrum of a substitution operator is closed under multiplication. From $\Omega(uv) = \Omega u \Omega v$ it follows that

(2.3) $\quad\quad \Omega(\phi_1 \phi_2) = \lambda_1 \lambda_2 \, \phi_1 \phi_2$ if $\Omega \phi_1 = \lambda_1 \phi_1$ and $\Omega \phi_2 = \lambda_2 \phi_2$

and consequently

(2.4) $\quad\quad \Omega(\phi^n) = \lambda^n \phi^n$ if $\Omega \phi = \lambda \phi$, $n = 1,2,3,\ldots$

The knowledge of one solution of the Schröder equation, therefore, leads to the knowledge of a countably infinite system of solutions, i.e. the powers of the first solution, with corresponding powers of the first eigenvalue as eigenvalues. Later, introducing what we shall term ξ-algebras, we shall be able to go farther in this direction.

Let us now formulate, with respect to substitution operators (2.2), and on the algebra P (see Sec. 1) of polynomials, the three principal questions outlined in Sec. 1.

I. The answer to the "reversal question" is in the affirmative. According to Lemma 1.1, on P, every operator of type H is (apart from the linear transformation $A\Omega - CI$) a substitution.

II. The answer to the solution question is negative. We would have to solve

(2.5) $\qquad p[q(z)] = \lambda p(z), \quad p,q \in P$

Let n be the degree of the given q and assume that there is a solution p with degree m. (2.5) then implies $mn = m$, excluding the trivial case $m = 0$ [i.e. $p(z) = $ const. and $\lambda = 1$ unless $p(z) \equiv 0$]. This means $n = 1$. A necessary (but, as we shall see, not sufficient) condition for the solvability of (2.5) in P is that q(z) be linear,

(2.6) $\qquad p(az + b) = \lambda p(z).$

As one sees easily, there is no solution for $a = 1$ (unless we consider the trivial case $p = $ const.), since $p(z + b) = \lambda p(z)$ implies $\lambda = 1$ and $p(z + b) = p(z)$ has no non-constant solution for $b \neq 0$. For $a \neq 1$, a solution is

(2.7) $$p_1(z) = z + \frac{b}{a-1}$$

as easily verified. According to (2.4), a family of solutions is given by

(2.8) $$p_n(z) = p_1(z)^n = \left(z + \frac{b}{a-1}\right)^n \text{ with } \lambda_n = \lambda_1^n.$$

III. The answer to the third principal question, the question of representation, can now be answered at once in the affirmative. Every element of can be represented as a linear combination of solutions p_n of the form (2.8).

Let us briefly comment on the deeper reason (beyond the evidence of the formal proof) why the answer to question I. is, in general, negative and why the Schröder equation has, in general, no solution in P. The solution of a functional equation is - in general - the result of a limiting process. It is therefore natural that in an algebra non-complete under a norm imposed on it, such as P, a functional equation should have in general no solution. As far as this plausibility argument is concerned, it is irrelevant that we did not impose any norm on P; neither do we have to investigate formally whether for a given algebra a norm exists under which it is complete.

Let us use (2.8) to solve a simple functional equation in a way analogous to a differential equation. (One of the basic aims of our approach is to emphasize the analogous role of substitution operators and first-order differential operators as special cases of Bourlet operators.)

The equation is

(2.9) $$p(az + b) - \alpha p(z) = s(z).$$

It is a special case of the Kordylewski-Kuczma equation (cf. [3]). The constants $a \neq 0, \neq 1$, and α are given and $s(z)$ is a given polynomial. We expand in terms of eigensolutions, i.e. solutions of the homogeneous equation, the Schröder equation

(2.10) $\qquad p(az + b) - \alpha p(z) = 0,$

and obtain

(2.11) $\qquad s(z) = \sum_{k=0}^{n} A_k \left(z + \frac{b}{a-1}\right)^k = \sum_{k=0}^{n} A_k p_k(z)$

(2.12) $\qquad p(z) = \sum_{k=0}^{n} B_k \left(z + \frac{b}{a-1}\right)^k = \sum_{k=0}^{n} B_k p_k(z)$

The A_k coefficients are known and can be determined, e.g., from the formula

(2.13) $\qquad A_k = \frac{1}{k!} s^{(k)}\left(\frac{b}{1-a}\right).$

Comparison of coefficients yields, because of

(2.14) $\qquad p_k(az + b) = a^k p_k(z) \qquad\qquad$ cf. (2.8)

the following solution for (2.9)

(2.15) $\qquad p(z) = \sum_{k=0}^{n} \frac{s^{(k)}\left(\frac{b}{1-a}\right)}{k!(a^k - \alpha)} \left(z + \frac{b}{a-1}\right)^k, \; \alpha \neq a^k, \; k = 0,1,2,\ldots,n.$

After this application of the method to a simple functional equation, let us proceed by considering the three principal questions for type H in the general form. So far we have considered the specific from (2.1) and even that on a special function algebra, namely P.

For simplicity we consider the case $A = 1$, i.e. $\Omega(uv) = (\Omega u)(\Omega v)$. We consider two sets, S_1 and S_2. We need not make any specification about their nature. Consider now a family of mappings $\{f\}$

$$S_1 \xrightarrow{f} S_2.$$

These mappings shall form a commutative linear algebra F. (Examples: 1. If S_2 is the set of complex numbers, addition, multiplication and scalar multiplication are the usual point-wise operations defined for functions. 2. If S_1 and S_2 are linear spaces, $\{f\}$ is a family of commuting linear mappings from S_1 to S_2 and addition, multiplication and scalar multiplication are defined in the operator sense. While example 1 is the subject of our present discussion, we come back to example 2, in a generalized form, when we deal with integral operators and their kernels.)

Now introduce linear mappings Ω on the algebra F, with the property $\Omega(uv) = \Omega u \Omega v$. Then the first principal question is this setting is the following. Is every such mapping Ω a "substitution", i.e. does there exist a mapping ω mapping S_1 into itself (<u>not</u> into S_2) such that

(2.16) $\quad\quad (\Omega f)(S_1) = f[\omega(S_1)]$

for every element $S_1 \in S_1$ and every $f \in F$? If yes, we have a mapping

(2.17) $\quad\quad \{S_1\} \xrightarrow{\omega} \{S_1\} \xrightarrow{f} \{S_2\}$, or

$\quad\quad\quad\quad \{S_1\} \xrightarrow{f \circ \omega} \{S_2\}.$

The second principal question can be formulated in the general way as follows. For a given linear endomorphism Ω on F, does there exist an element $\phi \in F$ such that $\Omega\phi = \lambda\phi$ holds? If the answer to the first principal question is affirmative, the formulation of the second question can be made more specific in the following. Does there exist a mapping ω of S_1 into itself, such that

$$\phi \circ \omega = \lambda\phi?$$

(Note that ϕ maps S_1 into S_2).

A number of results can be obtained even in this very general setting. A study of the proof of Lemma 1.1, for instance, convinces us that the answer to the first "principal question" is affirmative, if every element of f can be expressed by a given element f_o through a formal power series:

$$f = \sum_{n=o}^{\infty} A_n f_o^n.$$

Moreover, and even more generally, if the solution of the Schröder equation exists (even in the general sense of the "first form"), then, by a reasoning corresponding to (2.4), the mapping ϕ^n, $n = 2,3,4,\ldots$, is also a solution with eigenvalue λ^n.

Let us add a few remarks about invariants, i.e. solutions of the Schröder equations with $\lambda = 1$. First let $\phi(x)$ be a solution of the Schröder equation with eigenvalue λ. Then a system of invariants is given by

(2.17) $$\psi(x) = \tau\left[\frac{\log|\phi(x)|}{\log|\lambda|}\right].$$

Here $\tau(x)$ is any periodic function with period 1. Of course, one has to assume that all functions occuring belong to the algebra in question.

Let us mention, without proof, two theorems about invariants.

1. Let $\psi(x)$ be an invariant under substitution, and $w(z)$ a function holomorphic at and around $z = 0$. Then there exists a domain in which $w[\psi(x)]$ is an invariant. (Again one has to assume that the occuring functions belong to the algebra in question. The relevant item in the proof is that $w[\psi(x)]$ can be expanded into a power series in $\psi(x)$). The condition is sufficient: it is not known whether it is necessary.

2. Consider the algebra A of power series

$$\sum_{m,n=0}^{\infty} a_{mn} x^m y^n$$

with complex coefficients, convergent for all values $x^2 + y^2 \leq 1$ of the real variables x and y. Then the sub-set of A, the elements of which are invariant under the linear substitution

$$\begin{pmatrix} x \\ y \end{pmatrix} \to \begin{pmatrix} \frac{1}{2} & \frac{i}{2} \\ \frac{-i}{2} & \frac{1}{2} \end{pmatrix} \begin{pmatrix} x \\ y \end{pmatrix},$$

is identical with the set of functions of $x + iy$ holomorphic in the closed unit circle. [33].

(This definition of holomorphic functions is "global" rather than "local". It is also, in a sense, not profound since the definition of the algebra A itself involves convergent power series, i.e. a kind of analyticity. Still it is interesting that in this context holomorphy appears as an invariance property.)

A discussion of the Schröder equation $\phi \circ h = \lambda \phi$, where h is a linear function will follow in Sec. 3. We continue with an extension of the third "principal question". It concerns expansion of an arbitrary function in terms of a continuous family of solutions of the Schröder equation. (A corresponding method for the case of differential operators is the expansion method (see e.g. [7]) as developed by Weyl and later by Ikebe.) As seen from (2.4), a countable family of solutions of the Schröder equation is given if one solution is given. The situation improves further if we introduce what we shall term ξ-algebras.

A linear algebra A of complex-valued functions $f(p)$ - we do not specify the nature of the set of independent variables $\{p\}$ - shall be called a ξ-algebra, if

(2.18) $\qquad |f(p)|^{\xi} \varepsilon$ A, provided $f(p) \varepsilon$ A, for every $\xi > 0$.

Examples: The algebra $C(0,1)$ of continuous functions is a ξ-algebra. The algebra P of polynomials is not a ξ-algebra. Now let ϕ be the solution of the Schröder equation $\phi \circ h = \lambda\phi$, $\lambda > 0$, $\lambda \neq 1$, in a ξ-algebra A_ξ of functions of one real variable. Then the third principal question can be formulated this way: It is possible to represent every $g(x) \varepsilon A_\xi$ in the form

(2.19) $\qquad g(x) = \sum_{n=o}^{\infty} a_n \phi(x)^n + \int_{o}^{\infty} a(\xi)|\phi(x)|^{\xi} d\xi.$

A solution of the first-order functional equation

(2.20) $\qquad f[h(x)] - \alpha f(x) = g(x),$

(cf. the special case (2.9)), can now be found, if one solution of the

Schröder equation

(2.21) $\phi[h(x)] = \lambda\phi(x)$

is known. The procedure is analogous to that used in the explicit solution of (2.9), except that now the representation for the unknown f and the known g is of the form (2.19).

Note also that in a ξ-algebra the following statement is true as a particular case of the defining relation (2.18). $|\phi(x)|^\xi$ is a solution of a Schröder equation with eigenvalue $|\lambda|^\xi$ if ϕ is a solution with eigenvalue λ.

A direct consequence is the following. Given a solution of the Schröder equation with eigenvalue λ_o, $|\lambda_o| > 1$, it follows that all positive real numbers $\lambda > 1$, are eigenvalues belonging to some solution, since every such number can be represented in the form $|\lambda_o|^\xi$, for some $\xi > 0$. Therefore, introducing, on a normed algebra, the operator norm in the usual way, we can say that a substitution operator on such an algebra is unbounded if it has an eigenvalue the absolute value of which exceeds one. Given now a solution of the Schröder equation with eigenvalue λ_1, $|\lambda_1| < 1$, then, by a similar argument, every real number λ, with $0 < \lambda < 1$, is an eigenvalue. Moreover, the inverse of the operator, if it exists, is not bounded, since its spectrum contains the values $1/\lambda$.

In other words, if the spectrum of an operator contains any eigenvalue the absolute value of which differs from zero, then either the operator, or its inverse, or both, are unbounded. We note that the arguments given apply also to a non-ξ algebra, with the following modification. The spectrum of an operator with an eigenvalue λ_o, $|\lambda_o| > 1$, contains now a sequence, λ^n, which tends to infinity. For an eigenvalue λ_1, $|\lambda_1| < 1$, the corresponding sequence tends to zero, and our previous arguments are valid.

We summarize our findings using the term unitary in this context for operators the spectrum of which lies entirely on the unit circle.

Theorem 2.1. **Let** H **be a substitution operator on a normed function algebra and** H^{-1} **its inverse. Then** H **and** H^{-1} **are both bounded if and only if** U **is unitary.**

3. CONTINUOUS ITERATION. COMMUTING SUBSTITUTION OPERATORS

Historically, the Schröder equation emerged not, as in our presentation, as the eigenvalue equation of a substitution operator, but in a different, but related, context, the question of continuous iteration.

One defines the iterates $f_n(x)$ of a function $f(x)$ by

(3.1) $\quad f_n(x) = f[f_{n-1}(x)], \quad f_0(x) \equiv x.$

f_n is defined without limitation if the range of f is contained in the domain of f. If a unique inverse of f exists, we shall denote it by $f_{-1}(x)$, and the n-th iterate of it by $f_{-n}(x)$. The problem of continuous iteration is this. Let us find a function $F(x,\nu)$ with the properties

(3.2) $\quad F(x,n) = f_n(x), \quad n = 0,1,2,\ldots$

(3.3) $\quad F[F(X,\nu),\mu] = F(x,\mu+\nu].$

If (3.3) (the "translation equation", cf. [16], pp. 170-175) has a solution, we can put

(3.4) $\quad F(x,\nu) = f_\nu(x)$

and consider this function as the ν-th iterate of $f(x)$, where ν is a real, or even complex parameter. It is easy to find a solution for (3.3). Let $\phi(x)$ be a strictly monotonic function of one real variable, defined for all values of x, and let $\lambda \neq 1$ be a positive number. Then

(3.5) $\quad \phi_{-1}[\lambda^\nu \phi(x)] = F(x,\nu)$

is a solution of (3.3), as can be verified by direct calculation. Of course, (3.2) is not yet satisfied. For n = 1, we obtain the condition

(3.6) $$\phi_{-1}[\lambda\phi(x)] = f(x), \quad \text{or}$$

$$\phi[f(x)] = \lambda\phi(x),$$

i.e. precisely the Schröder equation. If this is satisfied, one sees from (3.6) and (3.5) that (3.2) is satisfied also for $n = 2,3,\ldots$.

Let us now discuss two solution methods. If f is a strictly monotonic function, defined for all x, we can form the sum

(3.7) $$\phi(x) \sim \sum_{n=-\infty}^{\infty} \lambda^{-n} f_n(x).$$

If this is convergent and if its sum belongs to the particular algebra in question, one sees by direct substitution that (3.7) is a solution of the Schröder equation. A suitable algebra to use this method is the algebra of bounded, continuous real functions defined for all x with norm

(3.8) $$\|f(x)\| = \underset{-\infty < x < \infty}{\text{l·u·b}} f(x),$$

(Cf. [17], p. 49.)

The question of the convergence of (3.7) is, of course, not yet settled by this. The situation becomes easier if one is searching for solutions with eigenvalue $\lambda = 1$, i.e. invariants. A solution of the invariant equation

(3.9) $$\Psi[f(x)] = \Psi(x)$$

is given by any convergent expression symmetric in $\{f_n(x)\}$, in particular

(3.10) $$\Psi(x) = \sum_{n=-\infty}^{\infty} \rho[f_n(x)]$$

where ρ is an arbitrary function. To show that a suitable choice of ρ ensures convergence, let us consider a strictly increasing, continuous function $f(x)$ with the following properties. (It will not form part of the algebra of bounded functions.)

(3.11) $\qquad f(0) = 0, \quad |f(x)| \geq |x|, \quad \alpha > 1$

and

(3.12) $\qquad \rho(x) = \dfrac{\sin |x|}{1 + |x|}$.

Then (3.10) becomes

(3.13) $\qquad \Psi(x) = \displaystyle\sum_{n=-\infty}^{\infty} \dfrac{|\sin|f_n(x)||}{1+|f_n(x)|}$.

Now from (3.11)

(3.14) $\qquad |f_n(x)| \geq \alpha^n |x|,$ and consequently

$\qquad |f_{-n}(x)| \leq \alpha^{-n} |x|$

and thus (3.13) is majorized by the series

(3.15) $\qquad \displaystyle\sum_{n=-\infty}^{\infty} \dfrac{|\sin \alpha^n |x||}{1 + \alpha^n |x|}$.

The part for $n \geq 0$ of (3.15) converges because it is majorized by the geometric series $\displaystyle\sum_{n=0}^{\infty} \alpha^{-n}$; the part for $n < 0$ because it is majorized by the same series. The estimates are

$$n \geq 0: \quad \dfrac{|\sin \alpha^n |x||}{1 + \alpha^n |x|} \leq \dfrac{1}{1 + \alpha^n |x|} \leq \dfrac{1}{|x|} \alpha^{-n},$$

$$n < 0: \frac{|\sin \alpha^n |x||}{1 + \alpha^n |x|} \leq |\sin \alpha^n |x|| \leq \alpha^n |x|$$

for $x \neq 0$. For $x = 0$, (3.11) ensures trivial convergence. Thus, it is in general easier to obtain explicit solution methods for invariants than it is for eigenfunctions in general. Let us point also to the theorem about invariants in the form $w[\psi(x)]$, mentioned in Sec. 2.

We now proceed to a number of problems concerned with the factorization or "decomposition" of operators. Given an operator Ω, we can ask the following questions. Does there exist a number of "simple" operators, e.g. Bourlet operators $\Omega_1, \ldots, \Omega_k$ such that

(3.16) $\qquad \Omega = \Omega_1 \Omega_2 \ldots \Omega_k$?

(This is the "multiplicative" decomposition. The question of "additive" decomposition will be touched in Sec. 4.) One can ask, whether the factors in (3.16) commute, whether the decomposition is unique, perhaps apart from the order of the factors, etc.

In the class of operators of type M, factorization (3.16) means simply factorization of the function ω. Thus in E, the algebra of entire functions, we shall have the Weierstrass factorization theorem, in P simply the factorization of a polynomial. In connection with the class D, it is known that second-order differential operators can be written - after sutiable transformation - as the product of two Bourlet operators in the class D.

The class of products of the form $D_i D_j$ where $D_i = D(\omega_i, \lambda_i)$, we called the Sturm-Liouville class under the assumption that each element of at least a sub-class of the algebra concerned vanish at two fixed values, e.g. 0 and 1. We shall return to the Sturm-Liouville class in Sec. 4.

Now, however, we are interested in type H, i.e. $(H\psi)(x) = \psi[\omega(x)]$. We start with a theorem about $\omega \in P$ found by J. F. Ritt in 1922 [18]. We shall call a polynomial p composite, if two non-linear polynomials, p_1 and p_2 exist such that

$$p(x) = p_1[p_2(x)], \quad p = p_1 \circ p_2.$$

Otherwise, we call p a prime polynomial. We shall call two decompositions equivalent,

$$p = p_1 \circ p_2 \ldots p_n \sim q_1 \circ q_2 \circ \ldots \circ q_n,$$

if a sequence of first-degree polynomials $\lambda_1, \ldots, \lambda_{n-1}$ exists such that

(3.17) $\quad q_1 = p_1 \circ \lambda_1, \, q_2 = \lambda_1^{-1} \circ p_2 \circ \lambda_2, \ldots, q_n = \lambda_{n-1}^{-1} \circ p_n$

Two equivalent decompositions we shall not consider different.

Theorem 3.1 (J. F. Ritt). Any two decompositions of a polynomial contain the same number of polynomials. The degrees of the polynomials in one decomposition are the same as in the other, except perhaps for the order in which they occur. In our terminology, in the algebra of polynomials, every substitution operator can be factorized in exactly n prime (non-factorizable) operators of the same class, n being the same for every decomposition, etc.

The next question is: Do the factors of the decomposition commute? A simple random trial convinces us that two polynomial substitutions, in general, do not commute. We are facing here an interesting situation. In the class of type M, any two operators commute. In the class of type D, two operators do not commute except for the trivial case that ω_1 and ω_2 differ only in a constant factor (cf. Lemma 1.1). In the class $H(\omega,1,0)$, however, the situation is more interesting. Two operators do not, in general, commute but there are wide sub-classes of commuting operators, one of them posing interesting unsolved problems.

We now turn to commuting families of polynomials. Let us define such a family as a countable set of polynomials $\{p_n(x)\}$ such that for any two elements

$$p_n \circ p_m = p_n[p_m(x)] = p_m[p_n(x)] = p_m \circ p_n.$$

We notice that the Chebyshev polynomials of the first kind,

$$T_n(x) = \cos[n \arccos x],$$

form such a family due to their property

(3.18) $$T_n \circ T_m = T_{nm} = T_m \circ T_n.$$

We shall investigate generalized classes of such polynomials, but to do so, we need the idea of an addition theorem.

We say that the function $f(x)$ possesses an addition theorem if a function of two variables $R(x,y)$ exists, defined in both variables on the co-domain of $f(x)$ such that

(3.19) $$f(x + y) = R[f(x), f(y)],$$

for every x and y in the domain of f (cf. [16], pp. 52-62).

Let us remind ourselves that while (3.19) is formally analogous to a multiplication theorem as (1.1), it is much simpler since (1.1) is a relation between elements of an (abstract) Banach algebra. After considering a Banach algebra of functions and writing the independent variable of these functions explicitly in (1.1), the situation - though still far from being analogous - becomes somewhat more similar to (3.19).

The Chebyshev polynomials of the first kind have several interesting properties. From among these, we choose the relation (3.18). We shall define as a system of generalized Cehbyshev polynomials (cf. [19]), a system satisfying (3.18). In this connection, we shall discuss semi-groups of commuting substitutions in general, reflect upon the connection between addition theorems and generalized Chebyshev polynomials, and mention one or two unsolved problems. First of all, the definition of the Chebyshev polynomials can be written in the form

(3.20) $\arccos (T_n(x)) = n \arccos x,$

i.e., arccos x is an eigenfunction with eigenvalue $\lambda = n$ of the substitution operator

$$Hf(x) = f[T_n(x)].$$

In other words, it is a solution of the Schröder equation

$$\phi[T_n(x)] = \lambda\phi(x).$$

Consider now the iterates of a function (in our terminology, the powers of a substitution operator; this latter terminology is to be preferred, since it comes nearer to the essence of the matter. If we want to "switch back" to functions, it is sufficient to apply the operator to the function x: $H^n x = f_n(x)$. In this connection, however, we are facing the problem of giving an "algebraic characterization of the element "x" of a function algebra, e.g. through fox = xof = f, for all f.) It is clear that the system $\{f_n(x)\}$(cf. (3.1)) forms a semi-group under substitution with $f_o(x) = x$ as the unit element. The system obtained by removing the unit element from the semi-group is generated by $f_1(x) = f(x)$. If the "Translation Equation" (3.3) has a solution, e.g. in the form (3.5), the discrete semi-group $f_n(x)$ can be embedded in the continuous semi-group $f_\nu(x) = \phi_{-1}[\lambda^\nu \phi(x)]$.

To obtain a similar embedding for the system of generalized Chebyshev polynomials in question, we choose in (3.5)

(3.21) $\quad\quad\quad \lambda^\nu = \mu$.

Carrying out this transformation of iteration index, we now have a continuous semi-group in which the composition law for the parameter is multiplication, not addition. This is now in conformity with (3.18) and we can attempt to embed our discrete system of polynomials into this continuous semi-group. One essential feature, however, has been lost in carrying out the transformation in (3.21). While $\{f_n(x)\}$, $n = 1, 2, \ldots$, is generated by $f(x)$, $\{P_m(x)\}$ is certainly not generated by $P_1(x)$. The elements of $P_m(x)$ are not iterates of the first element. We shall see that the $\{P_m(x)\}$ are still "iterates of each other" in a wider sense.

We shall not pursue the interesting line of thought arising from the observation that $\{P_m(x)\}$ is generated by the sub-system $\{P_{\pi_n}(x)\}$ where $\pi_0 = 1$ and π_n the n-th prime. This line of reasoning is close to Ritt's investigations.

Our "multiplicative" semi-group now may be defined by

(3.22) $\qquad P_m(x) = \psi_{-1}[m\psi(x)]$

where $\psi(x)$ satisfies conditions necessary for $P_m(x)$ to be well-defined. Embedding is now accomplished by simply writing a continuous non-negative parameter μ instead of m. Such is the case for the Chebyshev polynomials themselves, the continuous semi-group is

$$P_\mu(x) = \cos[\mu \arccos x].$$

All elements of P_μ are, in a wider sense, iterates of each other. They are "generated" by the same function ψ.

The situation is analogous to the following. Let M be a non-negative, self-adjoint n × n matrix. Consider the semi-group $\{M^n\}$, n = 0,1,2,... . It can be embedded into a continuous semi-group by defining

(3.23) $\qquad M^\nu = U^* D^\nu U,$

where $D = U M U^*$ is the diagonalized form of M, and D^ν is a matrix which contains the ν-th power of the eigenvalues of M in the main diagonal, the other elements being zero.

<u>This analogy shows that the solution of the Schröder equation, in the case of functions of one variable, amounts to the "diagonalization" of a substitution operator.</u>

Let us now approach the matter from another direction. Assume that a system of polynomials is given by the one-step recursion

(3.24) $$P_{m+1}(x) = R[P_m(x),x].$$

In order to establish a relation with the case of (3.22), we first bring (3.22) to the form (3.24). Assume that $\psi_{-1}(x)$ satisfies an addition theorem

(3.25) $$\psi_{-1}(x+y) = Q[\psi_{-1}(x),\psi_{-1}(y)].$$

Then, from (3.22) and (3.25)

$$P_{m+1}(x) = \psi_{-1}[m\psi(x) + \psi(x)] = Q[\psi_{-1}(m\psi(x)),x] = Q[P_m(x),x],$$

i.e. $P_m(x)$ has a one-step recursion formula of the form (3.24) with $R = Q$.

As an example, for the Chebyshev polynomials the corresponding addition theorem is that of the cosine:

$$\psi_{-1}(x+y) = \psi_{-1}(x)\psi_{-1}(y) = [(1-\psi_{-1}(x)^2)(1-\psi_{-1}(y)^2)]^{1/2}.$$

Therefore, the recursion formula for the Chebyshev polynomials is

$$T_{m+1}(x) = xT_m(x) - [(1-T_m(x)^2)(1-x^2)]^{1/2}.$$

Thus, an addition theorem induces a one-step recursion. It is not, conversely, the case that a one-step recursion necessarily induces a nontrivial addition theorem. We can see this immediately by noting that a system defined in the manner of (3.24) is not, in general, commutative. (But assuming the existence of a function ψ_{-1} satisfying an addition theorem, commutativity follows at once from (3.22).) Thus, in general, a one-step recursion does not give rise to an addition theorem with monotonic solutions.

The situation then is the following. Starting from the recursion, we can set up the addition theorem equation

(3.26) $\quad \psi_{-1}(x+y) = R(\psi_{-1}(x),\psi_{-1}(y))$

only to find that no monotonic solution ψ_{-1} exists, in general, so that the representation (3.22) does not exist. For example, the recursion

$$P_{m+1}(x) = [P_m(x) + x]^2, \quad P_0(x) = x,$$

gives rise to a system of polynomials. The corresponding addition theorem, however,

$$\psi_{-1}(x+y) = [\psi_{-1}(x) + \psi_{-1}(y)]^2$$

has a constant as the only solution. Thus the representation (3.22) does not exist and we cannot claim that the system commutes. In fact, a simple trial shows that it does not. Note that we did not claim the existence of the representation (3.28) is a necessary condition for commutativity. We shall come back to this — unsolved — problem. One can, however, formulate the following positive result. [19]

Lemma 3.1. *In order that the functions* $\{f_m(x)\}$, *defined by the recursion*

(3.27) $\quad f_{m+1}(x) = R[f_m(x),x], \quad f_0(x) = x,$

should form an Abelian semi-group under composition, it is sufficient that the functional equation (addition theorem)

(3.28) $\quad \sigma(x+y) = R[\sigma(x),\sigma(y)]$

should have a strictly monotonic solution.

Proof: If such a solution exists, we can define $\psi(x) = \sigma_{-1}(x)$ and form the continuous Abelian semi-group

(3.29) $\qquad g_\mu(x) = \psi_{-1}[\mu\psi(x)]$.

For $\mu = 1,2,3,\ldots$ we obtain the discrete Abelian semi-group $\{g_m(x)\}$, embedded in (3.29). Moreover, $P_o(x) = x$ and from (3.28) it follows that (3.27) is satisfied for $\{g_m(x)\}$. Since (3.27) is a one-step recursion and from (3.27) and (3.29) $f_o(x) = g_o(x) = x$, one finds $f_m(x) = g_m(x)$. This completes the proof.

In order that a semi-group of the type described be a system of generalized Chebyshev polynomials, each $f_m(x)$ must of course be a polynomial. Since the initial element $f_o(x) = x$, a necessary and sufficient condition should be contained in the properties of $R(x,y)$ in (3.28). An obvious sufficient condition $R(x,y)$ should be a polynomial in x and y. Surprisingly, this condition leads to essentially only one type of generalized Chebyshev polynomials. It is known (cf. [16]) that the only polynomial addition theorems admitting non-constant solutions are

(3.30) $\qquad \sigma(x+y) = \sigma(x) + \sigma(y) + C, \quad \sigma(x) = ax + C$, and

(3.31) $\qquad \sigma(x+y) = A\sigma(x)\sigma(y) + B[\sigma(x)+\sigma(y)] + \frac{B(B-1)}{A}$,

$\qquad\qquad \sigma(x) = \frac{1}{A}[\exp Cx - B]$.

While (3.30) gives rise to a system of linear functions, (3.31) leads to the system

$$P_m(x) = \frac{1}{A}[(Ax + B)^m - B]$$

with the one-step recursion

(3.33) $$P_{m+1}(x) = (Ax + B)P_m(x) + B\left(x + \frac{B-1}{A}\right).$$

(3.32) is the only non-trivial system of generalized Chebyshev polynomials derived from a polynomial addition theorem.

We now make a remark about fixed points. (3.29) can be written in the form(substituting $\psi_{-1}(x)$ for x)

(3.34) $$g_\mu[\psi_{-1}(x)] = \psi_{-1}[\mu x].$$

Writing $g_\mu(0) = \alpha$, this becomes

(3.35) $$g_\mu(\alpha) = \alpha$$

for all μ, and α is independent of μ. In other words, all elements of an Abelian semi-group of type (3.29) have a common fixed point. In Sec. 13, we shall point to some unsolved problems connected with these problems.

Concluding this section, we quote a relevant result by Ritt:

Lemma 3.2 (J. F. Ritt): *Let* $P(x)$ *and* $Q(x)$ *be two commuting polynomials not derived from the multiplication theorems of* exp x *or* cos x. *Then there exists a linear function* $\ell(x)$ *and a polynomial* $G(x) = xR(x^r)$, $R(x)$ *a polynomial, such that*

$$P(x) = \ell^{-1}[\varepsilon_1 G_m(\ell(x))]$$
$$Q(x) = \ell^{-1}[\varepsilon_2 G_n(\ell(x))]$$

where ε_1 *and* ε_2 *are r-th roots of unity.*

Ritt's result says that - apart from linear transformations - a system of commuting polynomials is generated by polynomial through iteration (composition with itself), except for the case "multiplication theorem of the exponential function" - in our terminology addition theorem (3.31) - and the "multiplication theorem of the cosine", in our terminology, the addition theorem of the cosine, leading to the Chebyshev polynomials proper.

4. RELATIONS BETWEEN BOURLET OPERATORS OF THE SPECIFIC FORM

Since the three types H, M, D or Bourlet operators all satisfy more or less special forms of the multiplication theorem (2.16), it is to be expected that there exist various simple relations between them. We shall search for such relationships in the specific form only (it would be more satisfactory to derive relations directly from the multiplication theorems), and only in the case of two special function algebras. The first of these is the algebra of entire functions of one real variable. We do not touch upon the problem of whether other Bourlet operators exist in this algebra besides those of the specific form.

Introducing the notations

(4.1) $\quad M(\omega)f = \omega(z)f(z)$

(4.2) $\quad H[\omega,A,\lambda]f = \frac{1}{A} f[A\omega(z)] + \lambda f(z)$

(4.3) $\quad D[\omega,\lambda]f = \omega(z) \frac{d}{dz} f(z) + \lambda f(z)$,

$f, \omega \in E$, where E is the algebra of entire functions and λ is a (complex) parameter. We now have

(4.4) $\quad D[\omega,0] = M(\omega)D[e,0]$,

where e is the unit element of E. Moreover we have in E

Lemma 4.1:

(4.5) $\quad H[\omega,1,0] = \sum_{n=0}^{\infty} \frac{M(\omega-z)^n D(e,0)^n}{n!}$.

To prove this, we follow C. Bourlet [11] and write in the Taylor expansion

$$f(b) = \sum_{n=0}^{\infty} \frac{f^{(n)}(a)}{n!} (b-a)^n, \quad b = \omega(z), \quad a = z,$$

i.e.

(4.6) $$f[\omega(z)] = \sum_{n=0}^{\infty} \frac{(\omega(z)-z)^n}{n!} \frac{d^n}{dz^n} f(z),$$

which is precisely (4.5).

Unfortunately, M and D do not commute except if M is a multiple of the identity. Therefore the r·h·s in (4.5) cannot, in general, be written as an exponential. (We shall see later that the connection between the class D and the class H has in a sense an exponential character.) In the case when M is a multiple of the identity, i.e. $\omega - z = h = $ const, we have

(4.7) $$H[z+h,1,0] = \sum_{n=0}^{\infty} \frac{[M(h) D(e,0)]^n}{n!} = \exp [M(h)D(e,0)],$$

expressing the fact

$$f(z+h) = \sum_{n=0}^{\infty} \frac{h^n}{n!} f^{(n)}(z),$$

and indicating that in this particular case a substitution operatos is the exponential of a differential operator. Choosing $h = 1$ and introducing the translation operator

(4.8) $$(Tf)(z) = H[z+1,1,0]f = f(z+1),$$

and writing for brevity $D(e,0) = D$, (4.7) becomes

(4.9) $\qquad T = \exp D.$

Now, we said that the classes S_D and S_H are in a sense exponentially related. (S_D denotes the class of operators of type D of the specific form on E etc.)

1. S_M is closed with respect to addition <u>and</u> multiplication,

(4.10) $\qquad M[\omega_1] + M[\omega_2] = M[\omega_1 + \omega_2],$

(4.11) $\qquad M[\omega_1]M[\omega_2] = M[\omega_1 \cdot \omega_2].$

i.e. multiplication in S_M is commutative.

2. S_D is closed with respect to addition, but not multiplication,

(4.12) $\qquad D[\omega_1, \lambda_1] + D[\omega_2, \lambda_2] = D[\omega_1 + \omega_2, \lambda_1 + \lambda_2].$

The product of two operators in S_D is not in S_D; it is not even a Bourlet operator. But we shall pay special attention to the Sturm-Liouville class S_{SL}

(4.13) $\qquad \Omega \in S_{SL},$ if $\Omega = D[\omega_1, 0]D[\omega_2, 0],$

i.e. $\qquad (\Omega f)(z) = \omega_1(z) \frac{d}{dz} \left[\omega_2(z) \frac{d}{dz} \right] f(z).$

Here $f \in E_0$, $E_0 \subset E$ being the algebra of entire functions which vanish at two fixed points, say 0 and 1. While multiplication leads out of the class S_D, formation of the commutator does not:

(4.14) $\qquad [D[\omega_1, \lambda_1], D[\omega_2, \lambda_2]] = D[w(\omega_1, \omega_2), 0],$

where $w(\omega_1, \omega_2)$ is the Wronskian determinant of ω_1 and ω_2. (4.14) suggests that the operators in S_D commute if and only if ω_1 and ω_2 are linearly dependent. In the case of two elements, this means that ω_2 is a scalar multiple of ω_1. (Incidentally, the identical vanishing of the Wronskian is necessary and sufficient for linear dependence, e.g. in the case of our algebra E of entire functions, but not for other algebras without qualification. See [12].) (4.14) can be obtained by straight-forward calculation; $D(0,0)$ is, of course, the zero operator.

3. S_H is closed under multiplication if $\lambda_1 = \lambda_2 = 0$.

(4.15) $\quad H[\omega_1,0] \cdot H[\omega_2,0] = H[\omega_2 \circ \omega_1, 0]$, where $\omega_2 \circ \omega_1 = \omega_2[\omega_1(z)]$.

We notice the fact that $\omega_1 \circ \omega_2 \in E$ whenever $\omega_1, \omega_2 \in E$. Thus E is one of the algebras in which all the three Bourlet operations can be carried out. This saves us a great deal of difficulties one encounters in other algebras. If λ_1 and λ_2 do not vanish, we obtain the generalization of (4.15):

(4.16) $\quad H[\omega_1,\lambda_1]H[\omega_2,\lambda_2] = H[\omega_2 \circ \omega_1, \lambda_1 \lambda_2] + \lambda_1 H[\omega_2,0] + \lambda_2 H[\omega_1,0]$.

In this general case S_H is not closed for multiplication. We see that S_D is closed for addition while at least an important sub-set of S_H is closed for multiplication according to (3.15). (The case $A \neq 1$ does not, in this connection, create any difficulty.) Thus is some vague sence S_H "contains the exponentials of the elements of T_D". A more precise meaning is attached to this vague statement in the special cases (4.7) and (4.9). We shall come back to this question.

Let us now investigate the inverses of Bourlet operators, if they exist. We shall carry through the steps first in the general setting in which we derived the multiplication theorems, i.e., Ψ shall be a commutative normed algebra with unit element and without divisors of zero, and Ω a Bourlet operator such that the inverse R, $R\Omega = I$, maps Ψ into itself. According to (1.16), applying R to both sides,

$$uv = AR[\Omega u \cdot \Omega v] + B[R(u\Omega v) + R(v\Omega u)] + CR(uv).$$

Since Ω has an inverse, we can introduce the new "variables"

$$\eta = \Omega u, \quad \xi = \Omega v, \quad \text{i.e. } u = R\eta \text{ and } v = R\xi$$

and obtain

(4.17) $\qquad R\eta R\xi = AR(\xi \cdot \eta) + BR[\xi R\eta + \eta R\xi] + CR[(R\eta)(R\xi)].$

This is not, of course, in general a multiplication theorem in the sense of (1.16) or even of (1.1). (Of course, as we saw, a multiplication theorem (1.1) is always of the form (1.16).) On the other hand, it resembles (1.5). Let us write in (4.17) $B = C = 0$, $A \neq 0$. We then obtain a multiplication theorem

(4.18) $\qquad R(\xi\eta) = \frac{1}{A} R\eta\, R\xi.$

Going back to the case of the function algebra E, (4.18) implies

(4.19) $\qquad H[\omega,1,0]^{-1} = H[\omega_{-1},1,0],$

where $\omega_{-1}(z)$ is the inverse function of $\omega(z)$. We do not enter the discussion of the difficulties obviously arising here and restrict ourselves to the remark that at least for the entire function $\omega(z) = az + b$, $a \neq 0$, the inverse function also is an entire function. Thus the set of operators (4.19) applies to is at least not empty.

We shall not investigate further the properties of the inverses of Bourlet operators. As already pointed out, the inverses of the operators in class D would lead us to a type of smoothing operators characterized by the Reynolds Identity (see (1.5)).

Let us add two remarks.

1. When we specialized our commutative algebra with unit element and with no divisors of zero to a function algebra, we always choose E, the algebra of entire functions and P the algebra of polynomials. Once we mentioned E_o, the algebra of entire functions vanishing at two distinct finite points. This latter shall be used in the discussion of Sturm-Liouville operators. This preoccupation with holomorphic functions should not be characteristic of the topic. We choose this solution to save space and time and still have something to demonstrate results on. Later we shall discuss an algebra of real functions.

2. We have been rather generous in imposing our conditions on the (abstract) algebra Ψ. A unit element can always be adjoined to an algebra without unit element. Introducing pairs of the form $[f,\alpha]$ (cf. [8], pp. 117-110), we then have

(4.20)
$$(f,\alpha) + (g,\beta) = (f + g, \alpha + \beta)$$
$$\lambda(f,\alpha) = (\lambda f, \lambda \alpha)$$
$$(f,\alpha)(g,\beta) = (fg + \alpha g + \beta f, \alpha \beta)$$
$$\|(f,\alpha)\| = \|f\| + |\alpha|,$$

then (0,1) is the unit element.

Also, we did not always use the condition that there are no divisors of zero. In fact, we used it only once, when treating the case $A \neq 0$. Thus our statements regarding the classes D and M are not dependent on the absence of divisors of zero.

In order to find exponential formulae more general than (4.9), we now consider the algebra R_1 consisting of differentiable functions defined for all x. (We do not impose any norm on R_1 and our considerations will not touch the question of completeness under any norm.) Consider the continuous one-parameter group of substitution operators H^ν on R_1 defined by

(4.21) $\quad (H^\nu f)(x) = f[h_\nu(x)],$ with $h_\nu(x) = \phi_{-1}[e^\nu \phi(x)].$

Here, $\phi(x)$ is a strictly increasing element of R_1 assuming all real values, i.e., $\lim_{x \to \pm\infty} \phi(x) = \pm\infty$.

Now define the generator G of H^ν by

(4.22) $\quad G = \lim_{\nu \to 0} \frac{H^\nu - I}{\nu}.$

After some calculation, one finds

(4.23) $\quad (Gf)(x) = \frac{\phi(x)}{\phi'(x)} \frac{d}{dx} f(x), \quad f \in R_1.$

Writing $H = \exp G$ this means heuristically

(4.24) $\quad H[h,1,0] = \exp D\left(\frac{\phi}{\phi'}, 0\right).$

This is a general exponential relation. The formula (4.24) is to be interpreted heuristically since we do not know whether the r·h·s is meaningful. In particular, one would need the condition $(\phi/\phi') \in C^\infty$, where C^∞ denotes the algebra of functions defined for all x and differentiable any number of times. (4.24) becomes a special case analogous to (4.7) if one puts $\phi = \alpha\phi'$, i.e.

$$\phi(x) = \exp\frac{x}{\alpha}, \quad h(x) = \phi_{-1}[e\phi(x)] = x + \alpha.$$

As opposed to (4.5), we now have an exponential formula

(4.25) $$f[h(x)] = \sum_{n=0}^{\infty} \frac{1}{n!}\left(\frac{\phi(x)}{\phi'(x)}\frac{d}{dx}\right)^n f(x)$$

Writing (4.24) in the form

(4.26) $$D(\frac{\phi}{\phi'},0) = \log H[h,1,0],$$

This formula leads to the following heuristic solution formula for the Schröder equation $\phi[h(x)] = e(x)$. For $f[h(x)] = (Hf)(x)$, we have

(4.27) $$\frac{\phi(x)}{\phi'(x)}\frac{d}{dx}f(x) = \sum_{n=1}^{\infty}\frac{(-1)^{n+1}}{n}(H-I)^n f(x)$$

provided the r·h·s converges in a suitable sense. For $f(x) = x$

(4.28) $$\frac{\phi(x)}{\phi'(x)} = \sum_{n=1}^{\infty}\frac{(-1)^{n+1}}{n}(H-I)^n x,$$

and we have found the following (needless to say, heuristic) solution for the Schröder equation:

(4.29) $$\phi(x) = \exp\left[\int_{x_0}^{x}\rho(t)dt\right]$$

with

$$(4.30) \quad \frac{1}{\rho(x)} = \sum_{n=1}^{\infty} \frac{(-1)^{n+1}}{n} \left(\sum_{k=0}^{n} \binom{n}{k} (-1)^k h_k(x) \right).$$

If in the Schröder equation e is replaced by some other $\lambda > 0$, one has e.g. in (4.29)

$$(4.31) \quad \phi(x) = \exp \left[\log \lambda \int_{x_0}^{x} \rho(t) dt \right]$$

while (4.30) remains unchanged. A formula of the type (4.30) emerges, in a different context, in the work of M. A. McKiernan. (See [13].) In McKiernan's paper, this series emerges in connection with the investigation of the Cayley-Schröder formula

$$(4.32) \quad h_\nu(x) = \sum_{n=0}^{\infty} \frac{\nu(\nu-1)\ldots(\nu-n+1)}{n!} \sum_{k=0}^{n} \binom{n}{k} (-1)^{n-k} h_k(x);$$

a sufficient condition is given for the convergence of (4.30).

There exists an alternative approach to (4.22). Let us assume that $h_\nu(x)$ as function of ν is differentiable from the right at $\nu = 0$. Then a calculation of $(Gf)(x)$, according to (4.22), leads after a straight-forward calculation to

$$(4.33) \quad (Gf)(x) = f'(x) \left(\frac{\partial h_\nu(x)}{\partial \nu} \right)_{\nu = +0}.$$

On the other hand, we have (4.23) and combining this with (4.33) we find

$$(4.34) \quad \frac{\phi(x)}{\phi'(x)} = \left(\frac{\partial h_\nu(x)}{\partial \nu} \right)_{\nu = +0}.$$

This formula is less interesting than it appears to be since, in general, $h_\nu(x)$ is not known without the explicit knowledge of $\phi(x)$. There are, however, cases in which $h_\nu(x)$ is easily found and, for the sake of completeness, we give the appropriate formulae. These can easily be derived from results in [5].

We consider the Moebius transformation with real variable and distinguish the following cases:

a) linear case $h(x) = ax + b$. We find

$$(4.35) \quad h_\nu(x) = \begin{cases} a^\nu x + \dfrac{a^\nu - 1}{a - 1} b, & a \neq 1 \\ \\ x + \nu b, & a = 1. \end{cases}$$

b) parabolic case: the fixed point equation $h(x) = x$ has a double root ξ_0. Writing

$$(4.36) \quad a = 1 + \beta\xi_0, \quad b = -\beta\xi_0^2, \quad c = \beta, \quad d = 1 - \beta\xi_0,$$

one finds

$$(4.37) \quad h_\nu(x) = \frac{(1 + \beta\xi_0\nu)x - \beta\xi_0^2\nu}{\beta\nu x + 1 - \beta\xi_0\nu}.$$

c) loxodromic case: the fixed point equation $h(x) = x$ has two (in general, conjugate complex) roots ξ_1, ξ_2. Writing

$$(4.38) \quad a = \xi_1\alpha - \xi_2, \quad b = -\xi_1\xi_2(\alpha-1)$$

$$c = \alpha - 1, \quad d = \xi_1 - \xi_2\cdot\alpha,$$

one finds

(4.39) $$h_\nu(x) = \frac{(\xi_1 \alpha^\nu - \xi_2)x - \xi_1 \xi_2(\alpha^\nu - 1)}{(\alpha^\nu - 1)x + \xi_1 - \xi_2 \alpha^\nu}.$$

Applying (4.34) to (4.35), (4.37) and (4.39), one finds the following solutions of the Schröder equation $\phi \circ h = e\phi$. (Solutions for $\phi \circ h = \lambda\phi$, $\lambda > 0$, $\lambda \neq 1$, can be obtained by taking the log λ-th power of the solution)

(4.40) $$\phi(x) = \begin{cases} \left(x + \dfrac{b}{a-1}\right)^{1/\log a} & , \ a \neq 1 \quad \text{(cf. (2.7) linear case)} \\ \exp \dfrac{x}{b} & , \ a = 1 \end{cases}$$

(4.41) $$\phi(x) = \exp\left[\frac{1}{\beta(x - \xi_o)}\right] \qquad \text{(parabolic case)}$$

(4.42) $$\phi(x) = \left|\frac{x - \xi_2}{x - \xi_1}\right|^{\frac{1}{\log \alpha}} \qquad \text{(loxodromic case)}$$

Invariants can be constructed using e.g. (2.17) and the theorem mentioned under 1. following (2.17).

We do not enter into a discussion of the periodicity properties and other properties of the function sequences $h_n(x)$, which can be derived from (4.37) and (4.39).

5. SUPERPOSITION OF SUBSTITUTION OPERATORS

The present section forms a bridge leading to the second part of our material – integral operators. There is no implication that integral operators in general (or even integral operators of any particular type) have to be introduced as superposition of substitution operators. The situation is simply that the present section belongs most properly here. The method we are going to use is essentially the same as in Section 3.

Let A be an algebra of real functions, defined on the real line, $\phi(x)$ differentiable and $\phi'(x) > 0$ for all x, $\phi(0) = 0$, $\lim_{x \to -\infty} \phi(x) = \infty$, $\lim_{x \to \infty} \phi(x) = -\infty$. Let $\phi \in A$ be then for a real λ

(5.1) $\qquad h(x) = \phi_{-1}[\lambda \phi(x)] \in A.$

We make the additional assumption

(5.2) $\qquad \lambda > 1.$

Under conditions (5.1) and (5.2), the ν-th iterate of h is given by (cf. (3.5))

$$h_\nu(x) = \phi_{-1}[\lambda^\nu \phi(x)],$$

ν being real. This enables us to introduce the continuous group $\{H^\nu\}$:

(5.3) $\qquad H^\nu f(x) = f[h_\nu(x)] = f[\phi_{-1}\lambda^\nu \phi(x)], \quad f \in A$

(5.3) enables us to introduce a superposition of substitution operators

(5.4) $\qquad \Omega f = \int_{-\infty}^{\infty} d\nu \rho(\nu) H^\nu f(x),$

where $\rho(\nu)$ is such that (5.4) converges whenever Ω is applied to any $f \in A$.

We are now going to show that (5.4) can be expressed as an integral operator of the conventional form

$$(5.5) \qquad (\Omega f)(x) = \int_a^b \omega(x,y)f(y)dy.$$

The proof consists of a simple verification. Using (5.3), (5.4) can be written as

$$(5.6) \qquad \Omega f = \int_{-\infty}^{\infty} \rho(\nu)f[\phi_{-1}(\lambda^\nu\phi(x))]d\nu.$$

Introducing the new variable

$$(5.7) \qquad \phi_{-1}[\lambda^\nu\phi(x)] = y$$

we find $\qquad \lambda^\nu\phi(x) = \phi(y)$

or $\qquad \phi(x)\lambda^\nu \log \lambda \, d\nu = \phi'(y)dy.$

Thus (5.6) becomes

$$(5.8) \qquad \Omega f = \int_a^b \rho\left[\frac{\log \phi(y) - \log \phi(x)}{\log \lambda}\right] \frac{f(y)}{\log \lambda} \frac{\phi'(y)}{\phi(y)} dy.$$

Taking into consideration, in short notation, $\phi(\infty) = \infty$ and $\phi(-\infty) = -\infty$, i.e. $\phi_{-1}(\infty) = \infty$ and $\phi_{-1}(-\infty) = -\infty$, we have, because of (5.2)

$$(5.9) \qquad b = \lim_{\nu \to \infty} \phi_{-1}[\lambda^\nu\phi(x)] = \infty \cdot \text{sgn } \phi(x)$$

$$(5.10) \qquad a = \lim_{\nu \to -\infty} \phi_{-1}[\lambda^\nu\phi(x)] = \phi_{-1}(0) = 0$$

We have proved

Lemma 5.1: *Let* $h(x) = \phi_{-1}[\lambda\phi(x)]$ *where* ϕ *is a differentiable, strictly increasing function which takes all real values;* $\phi(0) = 0$. *Then*

$$(5.11) \qquad \int_{-\infty}^{\infty} \rho(\nu) f[h_\nu(x)] d\nu = \int_{0}^{\infty \cdot \text{sgn } \phi(x)} \omega(x,y) f(y) dy$$

with

$$(5.12) \qquad \omega(x,y) = \frac{1}{\log \lambda} \rho\left[\frac{\log \phi(y) - \log \phi(x)}{\log \lambda}\right] \frac{\phi'(y)}{\phi(y)}$$

It is easy to verify that in (5.12) the argument of ρ is real, as it should be, for $\phi(y)/\phi(x)$ is never negative.

Consider now a few restrictions on ρ. If we postulate that ρ vanishes identically for negative ν, we are considering the semi-group $\{H^\nu\}$, $\nu \geq 0$, instead of the group $-\infty < \nu < \infty$. (5.11) then becomes

$$(5.13) \qquad \int_{0}^{\infty} \rho(\nu) f[h_\nu(x)] d\nu = \int_{x}^{\infty \text{ sgn } \phi(x)} \omega(x,y) f(y) dy$$

where we note sgn $\phi(x)$ = sgn x. Considering the case where ρ identically vanishes for positive ν, we have the semi-group $\{H^\nu\}$, $\nu < \infty$. (5.11) now becomes

$$(5.14) \qquad \int_{-\infty}^{0} \rho(\nu) f[h_\nu(x)] d\nu = \int_{0}^{x} \omega(x,y) f(y) dy.$$

Case (5.14) is the most interesting since it leads to an integral operator of the Volterra type. There is a seeming discrepancy between (5.13) and (5.14), the integration interval being infinite in the first case, finite in the second. This lack of symmetry is due to our choice $\lambda > 1$ (see (5.2)) as a consequence of which the "forward iterates" $h_n(x)$, $n \geq 0$, form a divergent sequence, while the sequence of inverse iterates, $n < 0$, converges to zero. Choosing $\lambda < 0$, cases (5.13) and (5.14) are reversed, a perhaps more elegant version.

Let us consider a few examples of superposition. The simplest case is that of the translation operator

(5.15) $\qquad Tf(x) = f(x-1)$.

We choose the negative sign for a reason which will soon become apparent. T is a substitution operator (linked to the first order differential operator $D = D(e,0)$ by (4.24).

Now $\phi(x) = \exp x$ with $\lambda = 1/e$. We however need not know ϕ^* since directly

(5.16) $\qquad h_\nu(x) = x - \nu$.

Superposition according to (5.4) results in

(5.17) $\qquad (\Omega f)(x) = \int_{-\infty}^{\infty} \rho(\nu) f(x-\nu) d\nu,$

or, after introducing the new variable $x-\nu = y$,

(5.18) $\qquad (\Omega f)(x) = \int_{-\infty}^{\infty} \rho(x-y) f(y) dy.$

We thus have found the Faltung-type (convolution-type) integral operators, characterized by the property that the kernel depends only on the difference of its variables. It appears as the result of the continuous superposition

---
*incidentally, $\phi \notin A$.

of translations.

The properties of the convolution are well known. Writing (5.18) in the form

$$\Omega f = \rho * f,$$

one can easily show that the operation $*$ is commutative. Perhaps, its most outstanding property is that under Laplace transformation, Faltung becomes multiplication:

(5.19) $\qquad L(\phi_1 * \phi_2) = L\phi_1 \cdot L\phi_2$

where $\qquad L\phi = \int_0^\infty e^{-xy} \phi(y) dy.$

This transform is used to transform linear differential operators with constant coefficients into multiplication operators, in our terminology, operators of type M.

6. INTEGRAL OPERATORS: REMARKS AND DEFINITIONS

As mentioned in the Introduction, we shall regard integral operators not so much as a class of operators as compared with differential, substitution operators, etc., but rather as a "standardized" way of representing operators.

In what ways can we construct integral operators?

1. We have seen that, in suitable function algebras, operators of the class D are first order differential operators. It follows that, certain conditions being satisfied, inverses of the operators in class D are integral operators, of a special kind.

2. Assuming we can embed the powers of an operator, $\{\Omega^n\}$, in a continuous semi-group $\{\Omega^\nu\}$, we can consider superpositions of the type $\int d\nu \alpha(\nu) \Omega^\nu$. We have seen in Sec. 5 that this approach leads to integral operators, if Ω is a suitable operator of class H.

3. We can consider families of linear functionals $F(\lambda)$ acting on the function space $L^2(a,b)$. If $a \leq \lambda \leq b$ and $F(\lambda)$ is a bounded functional for almost all values of λ, we have

(6.1) $\qquad F(\lambda)\phi(x) = \phi_\lambda$

and, according to the Riesz Lemma, there exists a fixed element $k_\lambda(x) \in L^2(a,b)$ such that

$$(6.2) \quad F(\lambda)\phi(x) = \int_a^b k_\lambda(x)\phi(x)dx = \Phi_\lambda.$$

Now treating λ as a variable, (6.2) becomes

$$(6.3) \quad F(\lambda)\phi(x) = \int_a^b k(\lambda,x)\phi(x)dx$$

with $k(\lambda,x) \in L^2$ for almost all λ or

$$(6.4) \quad \int_a^b |k(\lambda,x)|^2 dx < \infty \text{ for almost all } \lambda.$$

The class of operators described by (6.3) and (6.4) is the Carlemann class containing as an important sub-set the Hilbert-Schmidt class. We shall discuss these operators in detail.

It has to be added that in the manner described these operators are outside the framework of our approach in Sec. 1-5, since the space $L^2(a,b)$ cannot be made into an algebra using ordinary (pointwise) multiplication. If we want to work with point-wise multiplication, one way out is to consider a sub-class of $L^2(a,b)$, $-\infty < a < b < \infty$, such that the functions are bounded in absolute value. Such an algebra is non-complete. The same remark applies to the next point:

4. We can consider the "weighted average" (generalized arithmetic mean) of a function $\phi(x)$ given by

$$(6.5) \quad A\phi(x) = \int_a^b a(x)\phi(x)dx$$

If we introduce whole families of such "averaging functionals', (6.5) becomes

(6.6) $$A(\lambda)\phi(x) = \int_a^b a(\lambda,x)\phi(x)dx,$$

an arrangement formally similar to (6.3), only there the integral form was not primarily present and came about by use of the Riesz Lemma and here the integral form is inherent in the "averaging" approach. On the other hand, we cannot say anything about the properties of $a(\lambda,x)$. These properties are rather a matter of definition, while, in the approach of No. 3, (6.4) followed.

Our present approach is closely linked both with No. 1 (inverses of operators in class D) and the theory of the Reynolds operators. (See e.g. [4].)

5. The most fundamental approach to integral operators is the following. Consider a space S of functions defined on a set D such that a reproducing kernel exists. Such a kernel is a one-parameter family $k(x,y)$ such that

(6.7) $$k(x,y) \in S$$

for all fixed values of the parameter x, and

(6.8) $$\int_D k(x,y)f(y)dy = f(x), \qquad \text{for all } f \in S.$$

Let Ω be a linear operator defined on S and let us write

$$"(\Omega f)(x) = \int [\Omega k(x,y)]f(y)dy",$$

the quotation marks indicating that these formulae are to be interpreted heuristically. The difficulty of this general approach is

a) that the family of functions must be sufficiently "regular" to permit the existence of a reproducing kernel,

b) that the interchange of integral and operator must be admissible, and

c) that $k(x,y)$, as a function of x for fixed y, must belong to the same family of functions as $f(x)$ so that $\Omega k(x,y)$ is defined.

As far as the existence of the reproducing kernel is concerned, this depends on the "regularity" properties of the function space in question. After what has been said under No. 4, (6.5) can be interpreted as saying that the function value at x is in a highly generalized sense the weighted average of the function values all over D. Now, for a non-continuous function no such representation can exist, since the function values are not connected at all. Assuming that a reproducing kernel exists [see (6.8)], it would be sufficient to change the function value at x while leaving the other values unchanged. Such a manipulation would leave the integral unchanged and the equality would no longer hold. For a continuous function, the values of which are "locally connected", a quasi-reproducing kernel can be forced into the situation by introducing the "Dirac delta function"

$$f(x) = \int_{-\infty}^{\infty} \delta(x-y) f(y) dy.$$

(We are not now referring to the rigorous treatment of distributions.) The "weight function" $\delta(x-y)$ emphasizes very strongly the values at and near x and suppresses completely the others. Since, however, $\delta(x-y)$ is not a continuous function at all, it cannot be called a reproducing kernel in the sense of the definition given. It is then clear which classes of functions have reproducing kernels - those, the function values of which are related not only locally, as with continuous functions, but "in the large", i.e. holomorphic functions and - in the real - harmonic functions. (See e.g. [21].)

In what follows, we plan to single out approach No. 3 which leads to the theory of Carleman integral operators. Let $F(\lambda)$ be then a one-parameter family of linear functionals, defined on $L^2(a,b)$ and bounded for almost all λ. We demand also that $a \leq \lambda \leq b$. We then quote the lemma of F. Riesz, according to which every bounded linear functional on $L^2(a,b)$ can be represented by a scalar product. This leads to

(6.9) $$F(\lambda)f(x) = (\phi_\lambda, f) = \int_a^b \phi(\lambda,y)f(y)dy.$$

Since we assumed that $F(\lambda)$ is a linear bounded functional for almost all values of λ, we have

(6.10) $\qquad \phi(\lambda,y) \in L^2(a,b;y),$ \hfill for almost all λ.

The index "y" indicates that ϕ is an L^2-function as a function of y. Accordingly,

(6.11) $$\int_a^b |\phi(\lambda,y)|^2 dy < \infty, \qquad \text{for almost all } \lambda.$$

We can now forget about the way we introduced our operator and define a Carleman integral through

$$\text{(6.12)} \quad (Kf)(x) = \int_a^b k(x,y)f(y)dy, \qquad f \in L^2(a,b)$$

with

$$\text{(6.13)} \quad \int_a^b |k(x,y)|^2 dy < \infty, \qquad \text{for almost all } x.$$

These operators were introduced by T. Carleman in 1923,([22]) in a slightly different form, as far as additional conditions were imposed on the kernel beside (6.13). Also, only hermitean-symmetric kernels were included in the definition. Our approach, through a one-parameter family of linear functionals, can be justified by the argument that the "scalar product" quality of (6.13) appears to be the essential feature in treating these operators. The proof of our "Lemma of Right Multiplication", to be described, is based on this "scalar product" property of the operator.

Let us now define an important sub-class of the Carleman class, the (generalized) Hilbert-Schmidt class defined by

$$\text{(6.14)} \quad (\Omega f)(x) = \int_a^b \omega(x,y)f(y)dy, \qquad f \in L^2(a,b),$$

with

$$\text{(6.15)} \quad \int_a^b \int_a^b |\omega(x,y)|^2 dxdy < \infty.$$

It is equivalent to the property

$$\text{(6.16)} \quad \sum_n \|\Omega \phi_n\|^2 < \infty,$$

for every complete orthonormal set $\{\phi_n\}$ in L^2. The value of the sum in

(6.16) is independent of the choice of $\{\phi_n\}$. A generalized Hilbert-Schmidt kernel (generalized because we do not insist on hermitean symmetry) belongs to a space $L^2(a,b)$ of functions of two variables, and (6.15) defines its norm in this function space. The norm - we sometimes describe it as the "double norm" $|||\Omega|||$ of the operator Ω - is different from the operator norm $\|\Omega\|$. One has

(6.17) $\qquad \|\Omega\| \leq |||\Omega|||$

In the self-adjoint case Ω has a pure point spectrum $\{\lambda_n\}$ which forms a square convergent series and (6.17) can be visualized as

(6.18) $\qquad \|\Omega\| = |\lambda_{max}|, \qquad |||\Omega||| = \sum |\lambda_n|^2.$

The aim of the following sections is to investigate, not the solution of integral equations, but the representation of operators on L^2 as Carleman integral operators. Representability of an operator as a Carleman integral operator is a highly interesting property for various reasons. It is not unitarily invariant; K may be a Carleman integral operator while UKU* is not. If K is a Carleman integral operator, it is not trivially evident, although true for normal K, that K* is one; if K has the kernel $k(x,y)$, it does not follow that K* has the kernel $\overline{k(y,x)}$! These are only two of the peculiar features of the theory. One reason why the study of these operators is interesting is the following. Given an operator like (6.12), every item of information about the operator K is somehow contained in $k(x,y)$ and vice verse. Thus, properties defined in functional analysis can be traced back to properties defined in classical analysis in a straightforward manner. The results we are going to describe are based on work commenced by H. Weyl in 1909 [23] and continued by J. von Neumann in 1935 [24].

A new investigation of the problem was suggested by J. M. Jauch who came across it in connection with his work on the mathematical formulation of scattering theory in quantum physics [25]. Many of the results are contained in a 1963 paper by M. Misra, D. Speiser and myself [26]. M. Schreiber has done extensive work on Carleman integral operators contained in [27] and in research reports. (Schreiber uses in the paper quoted a different definition of a Carleman, resp. "Semi-Carleman" operator, which however coincides with ours in the self-adjoint case.)

Let us conclude by listing some of the questions we are going to answer. We shall present the Weyl-von Neumann results on the spectrum of Carleman integral operators. It shall turn out that zero must be a "Weyl limit point" of the spectrum of the operator. This, however, is only a necessary condition. As a preliminary, we shall touch upon Weyl's result stating that a continuous spectrum can be changed to a pure point spectrum (although one which few physicists would willingly accept as a point spectrum) by adding to the original operator an arbitrarily small perturbing operator which is complete continuous.

As a generalization of the Weyl-von Neumann results, we shall investigate the case of normal operators and in particular show that unitary operators are essentially unitarily equivalent to the sum of the identity and of a Carleman integral operator.

We shall proceed to consider strong Carleman integral operators, i.e. such operators for which the integral property is unitarily invariant. Reversing the known result that every Hilbert-Schmidt operator is strong, we shall demonstrate that a bounded, strong Carleman integral operator is necessarily a Hilbert-Schmidt operator — leaving open the problem whether unbounded strong Carleman integral operators exist or not.

Also we shall come across the connection between Carleman integral operator theory and the convergence almost everywhere of series of the form

(6.19) $$\sum_{n=0}^{\infty} a_n |\phi_n(x)|^2, \qquad a_n \geq 0,$$

where $\{\phi_n(x)\}$ is a complete orthonormal system.

Another of our problems shall be this. Given K with kernel $k(x,y)$ and assuming that UKU* is again a Carleman integral operator (as said, this is always true if K is a Hilbert-Schmidt operator), what is the explicit form of the kernel UKU*? In the above context, various other results and problems will be mentioned.

7. THE THEOREMS OF WEYL AND VON NEUMANN ON HERMITEAN CARLEMAN OPERATORS

We start with (see [24])

Theorem 7.1. (Weyl-von Neumann): Given any hermitean operator A on L^2, there exists a Hilbert-Schmidt operator X with arbitrarily small double-norm, such that A + X has a pure point spectrum; the set of limit points of the spectra of A and of A + X is identical.

The spectrum of the arbitrary (not necessarily bounded!) A will, in general, consist of a discrete part and of a continuous part. After the addition of X, the continuous part is replaced by a (countable!) set of discrete eigenvalues. Because of the last part of the Theorem just stated, these must form a dense set in each integral occupied by the continuous spectrum of A; otherwise the set of limit points would change. X is a Hilbert-Schmidt operator in the narrower sense, i.e. self-adjoint, thus A + X is self-adjoint, its spectrum is real, the eigenfunctions form a complete orthonormal set; thus, by definition of L^2, the discrete spectrum is countable.

Since we mentioned - see (6.17) - that the double norm is not smaller than the operator norm, the "correctional" operator has an arbitrarily small norm as well as an arbitrarily small double norm.

The proof of the Theorem - which we shall not give here - makes use of the fact, proved by Weyl [23], that addition of a completely continuous self-adjoint operator to an arbitrary self-adjoint operator does not change the set of limit points of the spectrum of the latter.

As far as Carleman operators are concerned, the importance of the Weyl-von Neumann theorem is very great; it reduces the study of self-adjoint operators with arbitrary spectrum to the study of self-adjoint operators with a (countable) discrete spectrum. The reduction of the problem, as seen, is carried through by adding a Hilbert-Schmidt operator. The crucial point is, as we shall see, that any Hilbert-Schmidt operator, as well as any unitary transform of it, is always a Carleman operator. The question, whether a given operator is unitarily equivalent to a Carleman operator or not, reduces to the investigation of the operator with a discrete spectrum.

Let us first prove

Theorem 7.2. (von Neumann): A (self-adjoint) Hilbert-Schmidt operator is a Carleman operator.

Proof: By definition, our Hilbert-Schmidt operator has a pure point spectrum $\{\lambda_n\}$ which is quadratically convergent. Let $\Omega \phi_n = \lambda_n \phi_n$, where $\{\phi_n\}$ is a complete orthonormal set, then $\{\phi_n(x)\phi_m(y)\}$, $n,m = 0,1,2,\ldots$, is a complete orthonormal set in the space $\underline{L}^2$ of functions of two variables.

Let us choose the sub-set $\{\phi_n(x)\phi_n(y)\}$ and attach to it the (square convergent) spectrum $\{\lambda_n\}$; then, by the Riesz-Fischer theorem, there exists an $\omega(x,y)$ in $\underline{L}^2$ such that $\{\lambda_n\}$, together with a countable infinity of zeros, forms the set of Fourier coefficients of ω with respect to $\{\phi_n(x)\phi_m(y)\}$. This ω can be shown to be the kernel representing Ω:

(7.1) $$(\Omega f)(x) = \int \omega(x,y) f(y) dy.$$

It is sufficient to see this for the complete orthonormal set $\{\phi_n\}$. In fact,

$$\Omega \phi_n = \int \omega(x,y) \phi_n(y) dy = g_n(x);$$

in order to expand $g_n(x)$ in terms of $\{\phi_n(x)\}$, we write

$$(g_n, \phi_m) = \int g_n(x) \phi_m(x) dx$$
$$= \int\int \omega(x,y) \phi_n(y) \phi_m(x) dx dy = \begin{cases} \lambda_n, & m = n \\ 0, & m \neq n \end{cases}.$$

Thus, $g_n(x) = \lambda_n \phi_n(x)$ and (7.1) is demonstrated to be correct. Let us remark that it is also usual to define Hilbert-Schmidt operators through (7.1).

Since the spectrum is unitarily invariant, i.e. Ω and $U\Omega U^*$ have the same spectrum, it follows that every unitary transform of a Hilbert-Schmidt operator is again such an operator. (We shall later come back to the question regarding the transformation which the kernel $\omega(x,y)$ undergoes when Ω is replaced by $U\Omega U^*$.)

We now come to the central problem: when is an operator on L^2 a Carleman operator? As far as we know, this property cannot be reduced to other known simple properties of the operator. Von Neumann therefore modified the question. Let us use the following definition. An operator is of Carleman type if there exists a unitary operator U, such that UAU* is a Carleman operator. In other words, an operator is of Carleman type if it is unitarily equivalent to a Carleman operator.

Let us also introduce the notion of a Weyl limit point: λ is a Weyl limit point of the spectrum, if (a) λ is an eigenvalue with multiplicity ∞, or (b) λ is a limit point of a sequence of eigenvalues, or (c) λ belongs to the continuous spectrum. Of course, b) and c) can occur together, etc.

We now formulate von Neumann's main result:

Theorem 7.3. (von Neumann): A self-adjoint operator is of Carleman type if and only if zero is the Weyl limit point of its spectrum.

For the proof, we again refer to [24]; however, we shall make some remarks. As mentioned, and as to be proven later, among the bounded Carleman operators the Hilbert-Schmidt operators are the only ones not to "break down" under any unitary transformation.

We shall simplify the von Neumann notation in order to demonstrate the underlying idea. We first point out that it is sufficient to carry out the construction in the space $L^2(-\infty,\infty)$. The "von Neumann transforms"

(7.2) $\qquad (N_1 f)(x) = x^{-1/2} f(\log x)$, and

(7.3) $\qquad (N_2 f)(x) = \frac{1}{1-x} f(\frac{x}{1-x})$

have the following properties (verified by straightforward, but somewhat lengthy, calculations): (1) they are linear, (2) they are unitary, (3) N_1 maps $L^2(-\infty,\infty)$ into $L^2(0,\infty)$ and N_2 maps $L^2(0,\infty)$ into $L^2(0,1)$, and (4) both transformations leave the Carleman property of an operator invariant. (The whole calculation is given in Appendix II to [26].)

Ultimately, the transformation

(7.4) $\qquad (N_3 f)(x) = \frac{1}{(b-a)^{1/2}} f(\frac{x-a}{b-a})$

provides a mapping between $L^2(0,1)$ and $L^2(a,b)$, a and b being finite. It is easily seen that N_3, too, possesses the properties (1) - (4) attributed to N_1 and N_2. It shall therefore be sufficient to investigate $L^2(-\infty,\infty)$.

The von Neumann construction is based on a careful "pairing" of eigenvalues $\{\lambda_n\}$ and eigenfunctions $\{\phi_n\}$ such that the expansion

(7.5) $$k(x,y) = \sum_n \lambda_n \phi_n(x) \phi_n(y)$$

converges in the pointwise sense and defines the kernel, which is then shown to have the Carleman property (6.13) and to represent the operator A. The procedure we called "careful pairing" fails if we replace $\{\phi_n\}$ by another complete orthonormal set or, possibly even if we permute $\{\phi_n\}$ in an essential way; this is the reason why a Carleman operator may become a non-Carleman operator through unitary transformation. (It remains, however, of Carleman type by definition.) In what follows, we shall describe newer developments of the theory based on the Weyl-von Neumann results.

8. RECENT RESULTS ON CARLEMAN OPERATORS

We now proceed to recent developments in the theory. The von Neumann theory of Carleman operators was developed for hermitean ($A = A^*$) operators; some aspects of this theory can be generalized to normal operators ($AA^* = A^*A$) and even more general operators. We simply define a Carleman operator by (6.13) without demanding hermiticity or normality. Some results will be specific for normal operators.

The first question we deal with is this: What can be said about products of the form KB and BK, where B is a bounded operator and K is a Carleman operator? While it will turn out that products of the BK are not, in general, Carleman operators, we have the following:

Lemma 8.1. ("Lemma of Right Multiplication") [26]: Let B be a bounded operator and K a (not necessarily bounded) Carleman operator; then KB is also a Carleman operator and, if K has the kernel $k(x,y)$, then KB has the kernel $CB^*Ck(x,y)$ where C stands for complex conjugation and B^* acts on functions of x and y as functions of y, while x is fixed.

The proof is contained in

$$(8.1) \qquad (KBf)(x) = \int_a^b k(x,y)(Bf)(y)dy$$

$$= (Bf, Ck_x) = (f, B^*Ck_x)$$

$$= \int_a^b f(y)[CB^*Ck(x,y)]dy.$$

We have to show that the new kernel satisfies the Carleman condition (6.13). In fact

$$\|CB^*Ck(x,y)\| = \|B^*Ck_x(y)\| \leq \|B^*\|\cdot\|Ck_x(y)\| = \|B^*\|\cdot\|k_x(y)\|.$$

Since B and thus B* is bounded, $\|B^*\| < \infty$, and K being a Carleman operator, by definition $\|k_x(y)\| < \infty$ for almost all x, thus the new kernel is a Carleman kernel. This complete the proof of Lemma 8.1. An immediate consequence of this "Lemma of Right Multiplication" is a similar statement about Carleman-type operators.

Lemma 8.2. [26]: _If_ B _is a bounded operator and_ A _is a Carleman-type operator, then_ AB _is also a Carleman-type operator_.

Proof: If A is a Carleman-type operator, then by definition there exists a unitary operator U such that K = UAU* is a Carleman operator. According to Lemma 8.1, multiplication from the right of K by the bounded operator UBU* results in a Carlemen-type operator K':

(8.2) $\qquad K' = KUBU^* = UAU^*UBU^* = UABU^*;$

AB is therefore unitarily equivalent to a Carleman operator, and is therefore itself a Carleman-type operator.

We now ask: can a unitary operator be a Carleman operator? We notice first that the (hermitean) identity operator I is not a Carleman operator, not even, of course, a Carleman-type operator, since $\lambda = 1$ is an eigenvalue with infinite multiplicity, and no other eigenvalues exist; thus $\lambda = 0$ is certainly not the Weyl limit point of its spectrum. We now can prove:

Lemma 8.3. [26]: <u>No unitary operator is a Carleman-type operator, and therefore no unitary operator is a Carleman operator.</u>

<u>Proof</u>: Assume some unitary operator were a Carleman-type operator. Then, U* being bounded, by Lemma 8.2 UU* = I would be a Carleman-type operator. This contradiction proves Lemma 8.3.

There exist of course unitary integral operators, e.g. on $L^2(-\infty,\infty)$ the Fourier operator

(8.3) $$(Fg)(x) = \frac{1}{(2\pi)^{1/2}} \int_{-\infty}^{\infty} e^{ixy} g(y) dy.$$

However, F is not a Carlemen operator, since $\int_{-\infty}^{\infty} |e^{ixy}|^2 dy = \infty$. On the other hand, we can define, on a finite interval

(8.4) $$(F'g)(x) = \frac{1}{(2\pi)^{1/2}} \int_{a}^{b} e^{ixy} g(y) dy.$$

Here the kernel satisfies the Carleman condition, it is even a Hilbert-Schmidt operator; but being a Hilbert-Schmidt operator, it is not unitary, since the relation

$$\sum_n \|F'\phi_n\|^2 < \infty \quad \text{implies} \quad \lim_{n \to \infty} \|F'\phi_n\| = 0,$$

where $\{\phi_n\}$ is a complete orthonormal set. This is incompatible with

$$\|F'\phi_n\| = \|\phi_n\| = 1.$$

To extend the von Neumann results to normal operators, we need the fact that every normal operator can be decomposed in a unique way into

the (commuting) product of a (non-negative) hermitean operator and a unitary
operator

(8.5) $\quad\quad\quad N = RU = UR.$

We state

Theorem 8.1. [26]: <u>A normal operator is a Carleman-type operator if and only if the hermitean factor in its polar decomposition is a Carleman-type operator. Therefore: a normal operator is a Carleman-type operator if and only if zero is the Weyl limit point of its spectrum.</u>

The sufficiency of the condition follows from Lemma 8.2; together with R, RU is also a Carleman-type operator. The necessity also follows easily: if RU is a Carleman-type operator, so is $RUU^* = R$. In a similar fashion we derive

Lemma 8.4. [26]: <u>A normal operator is a Carleman operator if and only if the hermitean factor in its polar decompsoition is a Carleman operator.</u>

One can combine Theorem 8.1 (the generalization of von Neumann's theorem) and Lemma 8.4 in the following statement:

<u>A normal operator N and its hermitean factor R are both Carleman-type operators, or neither of them is; and, within a class of unitarily equivalent normal Carleman-type operators, either both N and R are Carleman operators, or neither of them is.</u>

Using a terminology borrowed from mathematical physics, one could say that N and R are Carlemen operators "in the same representation".

Let us conclude this section by proving a negative result. While the sum of two Carleman operators is, by the Triangle Inequality, again a Carleman operator:

(8.6) $$\|k_1 + k_2\| \leq \|k_1\| + \|k_2\|,$$

it is not true that the sum of any two Carleman-type operators is a Carleman-type operator. To find a counterexample, decompose L^2 into two orthogonal subspaces $L^2 = L_1^2 + L_2^2$, L_1^2 spanned by $\{\phi_n\}$, L_2^2 by $\{\psi_m\}$; therefore $(\phi_n, \psi_m) = 0$ for all m,n. Consider now the two commuting hermitean operators A and B, both diagonal in the frame given by $\{\phi_m\}$ and $\{\psi_n\}$, and let the spectra be

A: $\quad \alpha_m^{(1)} = \frac{1}{m} \quad , \quad \alpha_n^{(2)} = 1 - \frac{1}{n}$

B: $\quad \beta_m^{(1)} = 1 - \frac{1}{m} \quad , \quad \beta_n^{(2)} = \frac{1}{n}$

Obviously, both A and B are Carleman-type operators, since $\{\alpha_m^{(1)}\}$ and $\{\beta_n^{(2)}\}$ tend to zero; but $A + B = I$, and I is, as shown, not a Carleman-type operator.

One can construct an even simpler example showing that the product of two Carleman-type operators is not necessarily a Carleman-type operator. Consider the commuting, hermitean operators A,B, their eigenvalues defined by (we start numbering by $n = 1$!):

$$A\phi_n = \begin{cases} n\,\phi_n & , \quad n \text{ even} \\ \frac{1}{n}\,\phi_n & , \quad n \text{ odd} \end{cases}$$

$$B\phi_n = \begin{cases} \frac{1}{n}\phi_n , & n \text{ even} \\ n\phi_n , & n \text{ odd} \end{cases} ;$$

then the product $AB = I$ is certainly not a Carleman-type operator. The fourth question — Is the product of two Carleman operators a Carleman operator? — has not been answered by such simple methods.

Let us now apply our results to the question of the integral representation of the resolvent. We define the resolvent as usual by

(8.7) $\qquad R(\lambda) = (A - \lambda I)^{-1}.$

Let us remind ourselves of the definition of the spectrum of an operator A; on L^2, λ belongs to the:

<u>Resolvent set</u> , if $R(\lambda)$ is bounded

<u>Continuous spectrum</u>, if $R(\lambda)$ is unbounded and defined on a dense set

<u>Residual spectrum</u> , if $R(\lambda)$ is unbounded and defined on a non-dense set

<u>Point spectrum</u> , if $R(\lambda)$ does not exist.

For the resolvent we have the Hilbert relation

(8.8) $\qquad R_\lambda - R_\mu = (\lambda - \mu) R_\lambda R_\mu = (\lambda - \mu) R_\mu R_\lambda .$

We now prove an unpublished result found by B. Misra.[*]

<u>Lemma 8.5.</u> (B. Misra): <u>Let $R(\lambda)$, the resolvent of A, be a Carleman operator for one value of the resolvent set. Then $R(\lambda)$ is a Carleman operator for all values in the resolvent set.</u>[**]

<u>Proof:</u> Let λ_0 be the fixed value in the resolvent set, so that $R(\lambda_0)$ is a

---

[*] communicated by his permission

[**] Related results were found by J. B. Miller [28].

Carleman operator. Then from (8.8), with $\mu = \lambda_o$,

(8.9) $\quad\quad\quad R_\lambda = R_{\lambda_o} + (\lambda - \lambda_o) R_{\lambda_o} R_\lambda$.

Now R_{λ_o} is a Carleman operator by assumption; R_λ is a bounded operator by definition, since λ is in the resolvent set; thus $R_{\lambda_o} R_\lambda$ is a Carleman operator by Lemma 8.1 according to (8.9); R_λ is the sum of two Carleman operators and thus itself is a Carleman operator. If we re-formulate Lemma 8.5, saying Carleman-type operator instead of Carleman operator and useing Lemma 8.2 instead of Lemma 8.1, we have

Lemma 8.6: <u>Let the resolvent</u> $R(\lambda)$ <u>be a Carleman-type operator for one value of the resolvent set; then</u> $R(\lambda)$ <u>is a Carleman-type operator for all values of the resolvent set</u>.

Among the two lemmas, it is Lemma 8.5 which is important, while Lemma 8.6 makes a rather trivial statement (the same is true, in a sense, for Lemma 8.1 and Lemma 8.2)for the following reason. If, e.g. for a self-adjoint operator H, the resolvent $R(\lambda) = (H - \lambda I)^{-1}$ is a Carleman-type operator, for some fixed λ_o, this means that $R(\lambda_o)$ has zero as the Weyl limit point, therefore $R(\lambda_o)^{-1} = H - \lambda_o I$ has $\pm \infty$ as Weyl limit point, therefore H itself is not bounded, therefore $H - \lambda I$ is not bounded for any λ, thus $R(\lambda)$ has a Weyl limit point at zero for any λ, thus $R(\lambda)$ is a Carleman-type operator for any λ. This heuristic reasoning shows that Lemma 8.6 is a rather commonplace statement as opposed to Lemma 8.5. In general, a theorem stating that a given operator is a Carleman operator conveys much more information than one stating that the operator is merely of Carleman-type.

We saw, from Lemma 8.3, that no unitary operator is a Carleman operator, therefore not even of Carleman-type. We can however prove

Theorem 8.2. [26]: *For every unitary operator* U *there exists at least one real number* γ, $0 \leq \gamma < 2\pi$, *such that*

(8.10) $\qquad U = e^{i\gamma}I + A,$

where A is a Carleman-type operator. Therefore: every unitary operator is unitarily equivalent to the sum of a Carleman operator and a multiple of the Identity.

Proof: The idea is to subtract from U a suitable multiple of the identity so that the resulting operator should have a Weyl limit point at zero and therefore be of Carleman-type. As known (cf. [29]), every unitary operator U can be written in the form

(8.11) $\qquad U = e^{iH},$

where H is a self-adjoint operator with spectrum entirely contained in $0 \leq \lambda < 2\pi$. According to the Bolzano-Weierstrass theorem, the spectrum of H must have at least one Weyl limit point, say at $\lambda = \gamma$; then the spectrum of U, located on the unit circle, must have a Weyl limit point at $e^{i\gamma}$. Since the unitary operators U and $e^{i\gamma}I$ commute, their difference has a spectrum with a Weyl limit point at zero, and is therefore a Carleman-type operator:

(8.12) $\qquad U - e^{i\gamma}I = A,$

where A is a Carleman-type operator. By definition, a unitary V exists such that VAV* is a Carleman operator. Thus

(8.13) $$VUV^* = e^{i\gamma}I + VAV^* = e^{i\gamma}I + K,$$

where K is a Carleman operator, and this proves the second part of Theorem 8.2.

We notice that Theorem 8.2 could have been proven in a much more general form! We do so now and add at once a statement which shall be important in the theory of what we shall call "strong" Carleman operators:

Theorem 8.3. [26]: <u>Every bounded normal operator is unitarily equivalent to the sum of a Carleman operator and a multiple of the identity. If more than one such decomposition exists, they must all occur "in different representations", i.e. the unitary equivalence must be generated by different unitary transformations. The (possibly non-countably infinite) number of decompositions equals the number of Weyl limit points of the spectrum.</u>

The first part of the theorem is proven exactly on the lines of Theorem 8.2, i.e. the spectrum of a bounded normal operator has at least one (finite)Weyl limit point, we subtract a corresponding multiple of the Identity, etc. The second part is obtained if we realize that, for every Weyl limit point in the spectrum, a subtraction leading to a Carleman-type operator can be carried out. Assume now that there exist two different decompositions "in the same representation", i.e. (cf. (8.13))

(8.14) $$VBV^* = \alpha I + K_1 \quad \text{and} \quad VBV^* = \beta I + K_2,$$

where α, β are different constants, K_1, K_2 different Carleman operators; subtraction then leads to

$$I = \frac{1}{\alpha-\beta}(K_2 - K_1),$$

i.e., the identity were representable as a Carleman operator, which we know is not the case. This completes the proof of Theorem 8.3.

Using the ideas developed, we can at once prove

Lemma 8.7: If the decomposition of an operator into the sum of a Carleman operator and a multiple of the Identity exists, then it is unique.

Proof: Assume that two different decompositions exist. Then

(8.15) $\quad A = \alpha I + K_1 \quad$ and $\quad A = \beta I + K_2$.

Here $\alpha \neq \beta$ and $K_1 \neq K_2$; for if $\alpha = \beta$, then K_1 and K_2 follows, and the two decompositions are identical; the same reasoning applies if $K_1 = K_2$. Subtractions leads to

$$I = \frac{1}{\alpha-\beta} (K_2 - K_1),$$

the same contradiction as in the case of the proof of Theorem 8.3.

Note that in Theorem 8.3 we had to make the assumption that the operator in question is bounded, in order to have at least one finite Weyl limit point, so that the decomposition can be carried through. In Lemma 8.7, we assumed the existence of a decomposition, thus boundedness has not to be assumed.

Let us add the remark, already anticipated when we discussed resolvent operators of Carleman-type, that the inverse of an unbounded normal operator, if it exists, is of Carleman-type. Let us conclude this section by

Lemma 8.8. [26]: If N is a normal (not necessarily bounded) Carleman operator, then the adjoint N* is also a Carleman operator.

Proof: Consider the polar decomposition (8.5) and note

(8.16) $\quad N^* = U^*R = RU^*.$

We now write (cf. (8.5))

$$N = RU$$

and multiply both sides by the (unitary, therefore bounded) operator $(U^*)^2$:

(8.17) $\quad NU^*U^* = RUU^*U^* = RU^*.$

By (8.16), (8.17) means

(8.18) $\quad N(U^*)^2 = N^*$

and by Lemma 8.1, $N(U^*)^2$ is a Carleman operator; therefore, so is N^*.

Note that the kernel $\nu^*(x,y)$ belonging to N^* is by no means necessarily identical with the kernel $\overline{\nu(y,x)}$. We come back to this problem in Section 13 (Problem No. VI).

9. STRONG CARLEMAN OPERATORS

Until now, we had to distinguish carefully between actual Carleman operators and operators which are unitarily equivalent to such operators, i.e. Carleman-type operators. Now we turn our attention to the class where we need not make this distinction.

<u>Definition</u>: We call a Carleman operator A strong, if UKU* <u>is a Carleman operator for every unitary U</u>.

It is obvious, that every unitary transform of a strong Carleman operator is again a strong Carleman operator. We commence our statements with a number of lemmas; some of these will be used merely to prove stronger results containing them.

<u>Lemma 9.1</u>. [26]: <u>If A is a strong Carleman operator and B is a bounded operator, then BA is a strong Carleman operator</u>.

<u>Proof</u>: It is known (cf. [30], p. 4) that every bounded operator can be written as a linear combination of four unitary operators

$$(9.1) \qquad B = \sum_{k=1}^{4} \alpha_k U_k.$$

Since the linear combination of any finite number of strong Carleman operators is easily seen to be a strong Carleman operator:

$$(9.2) \qquad U\left(\sum_{n=1}^{N} \alpha_n A_n\right) U^* = \sum_{n=1}^{N} \alpha_n U A_n U^*,$$

it will be sufficient to carry through the proof for unitary B; by definition, UAU* is a (strong) Carleman operator, and, therefore, multiplying from the right by U and applying Lemma 8.1, so is UA for any unitary U. Finally, UA is not only a Carleman operator, but a strong Carleman operator; to prove this, let us remind ourselves that U is an arbitrary unitary operator and we can replace it by VU, where U and V are arbitrary unitary operators; thus VUA is a Carleman operator, and, again applying Lemma 8.1, so is VUAV*, i.e., UA is a strong Carleman operator. Taking into consideration (9.1), this completes the proof.

An analogous statement is

Lemma 9.2. [26]: If A *is a strong Carleman operator and* B *is a bounded operator, then* AB *is a strong Carleman operator*.

Proof: It is again sufficient to treat the case B = U. Again, by definition, UBU* is strong and, by Lemma 9.1, U*UBU* = BU* is a Carleman operator. It is also strong, for, replacing U* by U*V* and applying Lemma 9.1 again by multiplying from the left by V, we find that VAU*V* is a Carleman operator, i.e., AU* is strong. Replacing U_k by U^*_k, in (9.1), we have completed the proof of Lemma 9.2.

The situation, to summarize, is then the following. For Carleman operators, right multiplication by bounded operators does not lead out of the class of Carleman operators, while left multiplication does. On the other hand, both right and left multiplication by bounded operators carries the class of strong Carleman operators into itself. We can formulate the latter statement as

Theorem 9.1. [26]: <u>The class of bounded strong Carleman operators is a two-sided ideal in the algebra of bounded operators on L^2</u>.

We now proceed to relate the class of strong Carleman operators to the class of Hilbert-Schmidt operators. We have quoted von Neumann's result (Theorem 7.2): a self-adjoint Hilbert-Schmidt operator is a Carleman operator. This result can be extended very easily. First of all, the Hilbert-Schmidt property is unitarily invariant; if A belongs to the Hilbert-Schmidt class, so does UAU* for arbitrary U. Thus, we can say: <u>A self-adjoint Hilbert-Schmidt operator is a strong Carleman operator</u>. Moreover, we can always decompose a given operator into a hermitean and an antihermitean part:

(9.3) $$A = \frac{A + A^*}{2} + i\frac{A - A^*}{2i},$$

and, as already remarked, the linear combination of a finite number of strong Carleman operators is again a strong Carleman operator. We can ultimately pronounce Theorem 7.2 in the form:

<u>A (not necessarily self-adjoint, or even normal) Hilbert-Schmidt operator is a strong Carleman operator.</u>

We now attack the reverse problem: Is every strong Carleman operator a Hilbert-Schmidt operator? We shall prove, through a number of steps, that every bounded, strong Carleman operator is a Hilbert-Schmidt operator. This will leave the unsolved problem, whether unbounded strong Carleman operators exist or not, and, if they exist, what can be said about them. For the sake of completeness, we formally state a result already discussed:

Lemma 9.3: A bounded Carleman operator is strong if and only if its hermitean and antihermitean parts are both strong.

The proof is immediate, once we know that A* is a strong Carleman operator, provided A is one. This result we take from

Lemma 9.4: Let A be a bounded, strong Carleman operator; then the adjoint A* is also a bounded, strong Carleman operator.

Proof: It is known (cf. [30], Chapter 1, § 6) that every two-sided ideal in the algebra of bounded operators on L^2 is closed under the adjoint operation. But Theorem 9.1 tells us that the class of bounded, strong Carleman operators is exactly such a class. This completes the proof.

We proceed by

Lemma 9.5: A bounded, strong, hermitean Carleman operator is completely continuous, therefore it has a discrete spectrum with zero as its only limit point.

Proof: The proof proceeds much on the lines of the proof of Theorem 8.3. First of all, if our operator A is strong, it has a Weyl limit point at zero. We are going to prove that there is no other Weyl limit point. Assume that such a second limit point, say $\lambda_o < \infty$ exists; then $A - \lambda_o I$ is a Carleman-type operator; there exists, by definition, a unitary U such that $U(A-\lambda_o I)U^*$ is a Carleman operator. A itself being strong, UAU^* is also a Carleman operator. Subtraction yields the contradiction that $\lambda_o I$, a multiple of the identity, is a Carleman operator; we know by Theorem 7.3 that this is impossible. Thus zero is the only finite limit point of the spectrum of A;

since A is bounded and thus ∞ cannot be a Weyl limit point, zero is the only one. (It is exactly here that the problem of the - possibly non-existant - unbounded strong Carleman operators arises. Certainly, from the above reasoning, we already know: if non-bounded strong Carleman operators exist, they must have exactly two Weyl limit points, one at zero and one at infinity. We think of the spectrum as a set of numbers in the complex plane; a Weyl limit point at infinity thus means that there is a point belonging to the spectrum outside any circle around zero.) A hermitean operator, the only Weyl limit point in the spectrum of which is zero, has necessarily a discrete point spectrum which moreover is countable (since L^2 is separable) and tends to zero. This proves Lemma 9.5.

We can now at once discard it for the stronger

Theorem 9.2. [26]: <u>A bounded operator on L^2 is a strong Carleman operator if and only if it is a Hilbert-Schmidt operator</u>.

The sufficiency of the condition is already contained in the generalized version of von Neumann's Theorem 7.2. To see the necessity, we go back to the self-adjoint case dealt with by Lemma 9.5. Written out in detail, this lemma says that given a bounded, strong hermitean Carleman operator A and a complete orthonormal set $\{\phi_n\}$, there exists a unitary U, a sequence $\{\lambda_n\}$ tending to zero, and a Carleman kernel $k(x,y)$ (see (6.11)) such that

(9.4) $\qquad (UAU^*\phi_n)(x) = \lambda_n \phi_n(x)$

(9.5) $\qquad (UAU^*f)(x) = \int k(x,y)f(y)dy.$

Combining (9.5) and (9.4), one finds

(9.6) $$\int k(x,y) \phi_n(y) dy = \lambda_n \phi_n(x)$$

i.e. $\{\lambda_n \phi_n(x)\}$ is the sequence of Fourier coefficients of the L^2-function $k(x,y)$ (as a function of y) with respect to $\{\phi_n\}$ and therefore

(9.7) $$\sum_n |\lambda_n \phi_n(x)|^2 < \infty .$$

Since $\{\phi_n\}$ can be any complete orthonormal system $\{\phi_n\}$ in L^2, we choose in the case of $L^2(a,b)$ with finite a and b, the system

(9.8) $$\phi_n(x) = (b-a)^{-1/2} \exp \frac{2\pi i n x}{b-a}$$

and find from (9.7)

(9.9) $$\sum_n |\lambda_n|^2 < \infty .$$

We can now rid ourselves of the condition that a and b are finite by using the unitary von Neumann transforms N_1, N_2, N_3 (see (7.2), (7.3), and (7.4)) which do not affect the spectrum. Thus, A has a pure point spectrum with (9.9); it is therefore a (hermitean) Hilbert-Schmidt operator. Ultimately, we can dispose of the condition of hermiticity by using Lemma 9.3 and combining this with the fact that an operator is a Hilbert-Schmidt operator if and only if hermitean and antihermitean parts are Hilbert-Schmidt operators. This completes the proof.

10. CONVERGENCE THEOREMS

As indicated in Sec. 6 (see (6.19)), the theory of Carleman operators is connected with series of the form

$$\text{(10.1)} \qquad \sum_{n=0}^{\infty} a_n |\phi_n(x)|^2$$

where $a_n \geq 0$ and $\{\phi_n(x)\}$ is some orthonormal system in L^2. In fact, a series of this type emerged in the proof of Theorem 9.2 (cf. (9.7)). We shall now investigate the convergence problems of series like (10.1). We first consider two complete orthonormal systems $\{\phi_n(x)\}$ and $\{\psi_n(x)\}$ on $L^2(a,b)$, the Hilbert space of complex valued functions of one real variable, defined on the interval $[a,b]$ with $-\infty \leq a < b \leq \infty$.

Lemma 10.1: *For every non-negative real sequence $\{c_n\}$ with $\lim c_n = 0$, there exists in $L^2(a,b)$ a complete orthonormal system $\{\phi_n(x)\}$ such that*

$$\text{(10.2)} \qquad \sum_{n=0}^{\infty} c_n |\phi_n(x)|^2 < \infty,$$

for almost all x in $[a,b]$.

Proof: Since $\lim c_n = 0$, the set $\{(c_n)^{1/2}\}$ has a Weyl limit point at zero; moreover, being real, it can be considered as the spectrum of a family of unitarily equivalent Carleman-type operators. By definition, at least one member of this family is actually a Carleman operator. Denoting the corresponding kernel by $k(x,y)$ and the corresponding complete orthonormal system of eigenfunctions by $\{\phi_n\}$, we can write

(10.3) $$(K\phi_n)(x) = \int_a^b k(x,y)\phi_n(y)dy = c_n^{1/2}\phi_n(x).$$

Now $\{c_n^{1/2}\phi_n(x)\}$ can be considered, because of (10.3), as the sequence of Fourier coefficients of the function $\overline{k(x,y)}$ (as a function of y, for fixed x), with respect to the system $\{\phi_n(y)\}$. But $\overline{k(x,y)}$, being a Carleman kernel, is by definition an L^2-function of y for almost all fixed values of x. The sequence of its Fourier coefficients is, therefore, square convergent for almost all x and (10.2) follows. This completes the proof of Lemma 10.1.

Assume now that we have the same sequence $\{c_n\}$ but a different interval $[\hat{a},\hat{b}]$. It is then possible to find an appropriate system $\hat{\phi}_n(x)$ such that (10.2) holds, by using one of the von Neumann transforms (cf. (7.2), 7.3), and (7.4)). A converse statement can easily be made: if

(10.4) $$\sum_{n=o}^{\infty} |a_n\phi_n(x)|^2 < \infty$$

for almost all x in [a,b], then the operator Ω on $L^2(a,b)$, defined by

(10.5) $$(\Omega\phi_n)(x) = |a_n|^{1/2}\phi_n(x)$$

is a hermitean Carleman operator. For, then there exists a function of y, $\overline{\omega(x,y)}$, which is an L^2-function for almost all values of x, such that $\{|a_n|^{1/2}\phi_n(x)\}$ is the set of its Fourier coefficients, i.e.

(10.6) $$(\Omega\phi_n)(x) = \int_a^b \omega(x,y)\phi_n(y)dy = \phi_n(y).$$

(10.6) defines ω for all elements of L^2. Note that $\{|a_n|\}$, as opposed to

$\{c_n\}$ in Lemma 10.1, is not <u>required</u> to have zero as its lower limit. This rather follows from the convergence of x almost everywhere in (10.4). Namely, assume that zero is not the lower limit of $\{|a_n|\}$, then (10.6) defines a Carleman operator the spectrum of which does not contain zero as a Weyl limit point. This proves

<u>Lemma 10.2</u>: If $\sum_{n=0}^{\infty} |a_n \phi_n(x)|^2 < \infty$ <u>for almost every</u> x <u>in</u> [a,b], <u>and</u> $\{\phi_n(x)\}$ <u>is an orthonormal set in</u> $L^2(a,b)$, <u>then</u> $\lim |a_n| = 0$.

The following theorem is a rather general one; special cases will be derived in the form of lemmas. Again $\{\phi_n\}$ is a complete orthonormal system in $L^2(a,b)$ and $\sum_n c_n |\phi_n(x)|^2 < \infty$ for almost all x; $c_n \geq 0$; (it follows from Lemma 10.2 that $\lim c_n = 0$); $\{\psi_n(x)\}$ is another complete orthonormal system in $L^2(a,b)$ and the unitary operator U is defined by

(10.7) $\qquad U\phi_n = \psi_n$.

K is the Carleman operator defined by $K\phi_n = c_n^{1/2} \phi_n$.

<u>Theorem 10.1</u>: $\sum c_n |\psi_n(x)|^2 < \infty$ <u>for almost all</u> x, <u>if and only if</u> UK <u>is a Carleman operator</u>.

Proof: The condition is sufficient. For let UK be a Carleman operator; then, by Lemma 8.1, UKU* is also a Carleman operator. Therefore there exists a kernel $h(x,y) \in L^2(a,b;y)$ such that

(10.8) $\qquad (UKU^*f)(x) = (Hf)(x) = \int_a^b h(x,y)f(y)dy$.

By definition of K and U

(10.9) $$UKU^*\psi_n = c_n^{1/2}\psi_n.$$

(10.8) and (10.9) imply

(10.10) $$\int_a^b h(x,y)\psi_n(y)dy = c_n^{1/2}\psi_n(x).$$

$\{c_n^{1/2}\psi_n(x)\}$ is the set of Fourier coefficients of the function $\overline{h(x,y)}$ $\varepsilon\ L^2(a,b;y)$ with respect to $\{\psi_n(y)\}$; thus it is square convergent:

$$\sum_n c_n|\psi_n(x)|^2 < \infty.$$

The condition is also necessary. For let $\{c_n^{1/2}\psi_n(x)\}$ be square convergent; then it is the sequence of Fourier coefficients with respect to $\{\psi_n(y)\}$ of some function $\overline{h(x,y)}\ \varepsilon\ L^2(a,b;y)$, i.e.

$$\int_a^b h(x,y)\psi_n(y)dy = c_n^{1/2}\psi_n(x),$$

and this defines the Carleman operator

$$(Hf)(x) = \int_a^b k(x,y)f(y)dy, \qquad \text{for all } f\ \varepsilon\ L^2(a,b).$$

But from the definition of U and of K, $H = UKU^*$, and, using Lemma 8.1, $UKU^*U = UK$ is a Carleman operator. This completes the proof of Theorem 10.1.

To derive criteria for the applicability of this rather generally formulated theorem, we prove

<u>Lemma 10.3</u>: ("Decomposition Principle"). <u>UK is a Carleman operator if and only if a (not necessarily bounded) linear operator</u> A <u>exists such that</u> $(U-A)K$ <u>and</u> AK <u>are both Carleman operators.</u>

Proof: If UK is a Carleman operator, the lemma follows with A being the zero operator. If A exists, according to (8.6),

(10.11) $\quad (U-A)K + AK = UK$

is a Carleman operator.

The decomposition (10.11) is useful for the following reason: criteria referring to the term $(U-A)K$ involve $\{\phi_n\}$ and $\{\psi_n\}$, while AK may be independent of $\{\psi_n\}$. We shall see some applications of this decomposition method. In particular, UK is a Carleman operator if, in the decomposition of Lemma 10.3, A is bounded and commutes with K and U-A is a Carleman operator, since now AK = KA is a Carleman operator according to Lemma 8.1.

Before specializing this further, we state a negative result; it is of no advantage to make the seemingly obvious choice $A = K^m$, except for $m = 0$. $U-K^m$ should be a Carleman operator K_1, i.e.

(10.12) $\quad U = K^m + K_1.$

Since now K is bounded, it follows from Lemma 8.1 that for $m \geq 1$, $K^m = K \cdot K^{m-1}$ is a Carleman operator and, from (8.6) (see (10.12)), that U is a Carleman operator. But this contradicts Lemma 8.3. A decomposition (10.11) with $A = K^m$ is, of course, not shown to be useless by this argument if K is unbounded. Thus $A = I$ is a possible choice, and we formulate the decomposition at once in a general form and return then to the case $A = I$.

Now let $\{\phi_n\}$ and $\{\psi_n\}$ again be two complete orthonormal systems in $L^2(a,b)$, $\{\mu_n\}$ a bounded sequence of complex numbers, and $\{b_n\}$ a bounded non-negative real sequence with $\lim b_n = 0$, $b_n \leq M$.

Theorem 10.2: If

(10.13) $\quad \sum_n \|\psi_n - \mu_n \phi_n\|^2 < \infty$, for almost all x and

(10.14) $\quad \sum_n b_n |\phi_n(x)|^2 < \infty$, for almost all x

then $\quad \sum_n b_n |\psi_n(x)|^2 < \infty$, for almost all x.

Before proving Theorem 10.2, we remark that $\{\mu_n\}$ cannot be entirely arbitrary; e.g. it cannot contain zero infinitely many times, for then the sequence (10.13) will contain $\|\psi_n\| = 1$ infinitely many times and thus will diverge. More generally, $\{\mu_n\}$ cannot have a limit point at zero, for then the linear operator defined by

(10.15) $\quad B\phi_n = \mu_n \phi_n$

is an operator of Carleman type, and there exists a unitary operator V such that VBV* is a Carleman operator. In this case, from (10.13) and $U\phi_n = \psi_n$,

(10.16) $\quad \sum_n \|(U-B)\phi_n\|^2 < \infty$.

Therefore U-B is a Hilbert-Schmidt operator and so is $V(U-B)V^* = \Omega$. This means $VUV^* = VBV^* + \Omega$; but here the right hand side is a Carleman operator, being the sum of two Carleman operators, while the left hand side is unitary. This contradicts Lemma 8.3.

To prove Theorem 10.2, we remark that B as defined by (10.15) is a bounded operator which also by (10.15) commutes with K, since they have a common complete system of eigenfunctions; K is also bounded since by definition $K\phi_n = b_n^{1/2}\phi_n$. Thus, Lemma (10.3) applies. (See the remarks following Lemma 10.3.) If $\mu_n = 1$, for all n, we obtain a special case which also corresponds to $m = 0$ in $A = K^m$. We formulate this separately as

Lemma 10.4: *If, under the conditions of Theorem 10.2,*

$$\sum_n \|\psi_n - \phi_n\|^2 < \infty$$

and

$$\sum_n b_n |\phi_n(x)|^2 < \infty ,$$

then

$$\sum_n b_n |\psi_n(x)|^2 < \infty .$$

For Lemma 10.4, the following elegant elementary proof has been given by K. Tandori[*]. In $|\alpha + \beta|^2 \leq 2(|\alpha|^2 + |\beta|^2)$, choose $\alpha = b_n \phi_n(x)$ and $\beta = b_n(\psi_n(x) - \phi_n(x))$. On summation with respect to n one obtains, for every fixed x,

$$\sum_n b_n |\psi_n(x)|^2 \leq 2 \sum_n b_n |\phi_n(x)|^2 + b_n |\psi_n(x) - \phi_n(x)|^2$$

$$\leq 2 \sum_n b_n |\phi_n(x)|^2 + M \sum_n |\psi_n(x) - \phi_n(x)|^2.$$

From $\sum_n \|\psi_n - \phi_n\|^2 < \infty$, it follows by Levi's Theorem that $\sum_n |\psi_n(x) - \phi_n(x)|^2$ converges for almost all x, and $\sum_n b_n |\phi_n(x)|^2$ converges by assumption. This completes the proof.

---

[*] personal communication

11. TRANSFORMATION OF STRONG KERNELS

By definition, if K is a strong Carleman integral operator, then so is UKU* for every unitary operator U; also, Theorem 9.1 states that the class of bounded, strong Carleman integral operators is a two-sided ideal in the algebra of bounded operators. Actually, somewhat more can be said: if unbounded, strong Carleman integral operators exist, the class of not necessarily bounded, strong Carleman integral operators is closed under multiplication by arbitrary bounded operators from the right, or from the left.

We shall however restrict ourselves to the class of bounded, strong Carleman integral operators, i.e. the (generalized) Hilbert-Schmidt class (of in general not normal operators). The question we pose is the following. What (obviously linear) transformation does the kernel undergo, while the corresponding operator is multiplied by bounded operators from the left, or from the right, generally from both sides (this case including the case of unitary equivalence)? We introduce the notation k(X) for the kernel belonging to the Carleman operator X.

We answer this question in a more general form by

Lemma 11.1: If the not necessarily normal Hilbert-Schmidt operator Ω has the kernel $\omega(x,y)$, then the Hilbert-Schmidt operator $A\Omega B$, (where A and B are bounded operators) has the kernel

11.1 $CB^*ECAE\ \omega(x,y) = EACEB^*C\ \omega(x,y)$

where C denotes the (anti-linear) operation of complex conjugation,

E interchanges the variables x and y, and B* is understood to act on the function ω as a function of y, while x remains fixed.

As said, we shall denote the kernel belonging to some Carleman operator X by $k(X)$; we shall need the following relationships:

(11.2) $k(X^*) = ECk(X)$ [†]

(11.3) $k(XB) = CB^*Ck(x)$ (Lemma 8.4)

and, of course,

(11.4) $EC = CE;\quad E^2 = C^2 = I.$

We now obtain the proof of Lemma 11.1:

$$k(A\Omega B) = k[(A\Omega)B] = CB^*Ck(A\Omega) = CB^*Ck[(\Omega^*A^*)^*] = CB^*CCEk(\Omega^*A^*)$$

$$= CB^*ECA^{**}Ck(\Omega^*) = CB^*ECACECk(\Omega) = CB^*ECAE\omega(x,y).$$

Similarly,

$$k(A\Omega B) = k[A(\Omega B)] = k\{[(\Omega B)^*A^*]^*\} = ECk[(\Omega B)^*A^*] = ECCA^{**}Ck[(\Omega B)^*]$$

$$= EACECk(\Omega B) = EAECB^*Ck(\Omega) = EACEB^*C\omega(x,y).$$

This completes the proof. Incidentally, we have found the following:

Lemma 11.2: Let $\underline{L}^2(a,b)$ be the space of functions $f(x,y)$ with $\int_a^b \int_a^b |f(x,y)|^2 dxdy < \infty$, and A,B be bounded operators acting on the space

---

[†] This holds not only for Hilbert-Schmidt operators, but for other classes too; but it is for obvious reasons of asymmetry false for Carleman operators in general. (See [31], p. 227, on the "P-property" and cf. Lemma 8.4 and Problem VI, Sec. 13.)

$L^2(a,b)$ of functions of one variable; let $Ef(x,y) = f(y,x)$ and $Cf(x,y) = \overline{f(x,y)}$; then

(11.5) $\quad$ CBECAE = EACEBC.

We notice that the order of components on the two sides in (11.5) is reversed with respect to each other. A and B are understood to act on the elements of $\underline{L}^2$ as on functions of y, while x is fixed.

If we want to denote explicitly the variable which is non-fixed during transformation, we can write B_x, B_y, respectively, etc. We can dispose of the operation E and, e.g., (11.5) takes the form

(11.6) $\quad CB_y CA_x = A_x CB_y C$.

We can write this result in yet another form, for the sake of simplicitly formulated on a real Hilbert space.

Let A and B be bounded operators on $L^2(a,b)$, and let them operate on $\underline{L}^2(a,b)$, a real space of functions $f(x,y)$ of two variables, in such a manner that A acts on f as on a function of x, and B acts on f as on a function of y; then $(Af)(x,y) \in \underline{L}^2$, $(Bf)(x,y) \in \underline{L}^2$, and

(11.7) $\quad$ ABf = BAf.

This group of results can be related to a theorem by M. Eidelheit [32] about automorphisms of operator algebras. The explicit form of an inner automorphism of the algebra of Hilbert-Schmidt operators is given by (11.1), writing $A = B^{-1}$.

Ultimately, for the case of unitary equivalence, $A = B = U$, (11.1) becomes:

(11.8) $\quad K(U\Omega U^*) = (CU^*E)^2 k(\Omega) = (EU^*C)^2 k(\Omega)$.

12. OPERATORS OF LOCALLY BOUNDED RANGE

We conclude our discussion of Carleman operators by proving a lemma which links these to what we shall call operators of locally bounded range. Let $L^2(a,b)$ be a Hilbert space of (real, or complex valued) functions of one real variable x, $a \leq x \leq b$. Let A be a bounded linear operator defined on $L^2(a,b)$. We shall say that A is of locally bounded range, if a non-negative number M exists, independent of f and x, such that

(12.1) $\qquad |(Af)(x)| \leq M\|f\|$, $\qquad$ for all $f \in L^2(a,b)$

and for every $a \leq x \leq b$.

The class of operators with locally bounded range is certainly not empty, as the example of the zero operator shows. To construct a non-trivial example, we choose a function $\psi(x) \in L^2(a,b)$ with the property $\|\psi(x)\| = 1$, $|\psi(x)| \leq M$ and define A by

(12.2) $\qquad (Af)(x) = \int_a^b \psi(x)\overline{\psi(y)}f(y)dy.$

The operator A is now a projection operator on a one-dimensional subspace; it is a self-adjoint Hilbert-Schmidt operator. One has

(12.3) $\qquad |(Af)(x)| = |\psi(x) \int_a^b \overline{\psi}(y)f(y)dy| \leq |\psi(x)| \cdot \|\overline{\psi}\| \cdot \|f\|$

$\qquad\qquad\quad = |\psi(x)| \cdot \|f\| \leq M\|f\|.$

A is therefore an operator of locally bounded range. We now give a necessary condition for an operator to have a locally bounded range:

Lemma 12.1: *If an operator is of locally bounded range, then it is a Carleman operator; and if it is normal, zero is a Weyl limit point of its spectrum.*

Before proving this lemma, we point out that the condition cannot be sufficient; it is easy to construct a Carleman operator the range of which is not locally bounded. Proceeding on the lines of our previous example for the existence of a non-zero operator with locally bounded range, we replace $\psi(x)$ by an unbounded L^2-function $\psi(x)$; then all f which are non-orthogonal to ψ are transformed by A into multiples of ψ, i.e., locally unbounded functions. A however is now still a Hilbert-Schmidt operator, a fortiori a Carleman operator.

To obtain the proof of Lemma 12.1, we introduce the linear functional $L(x_0)$ by

(12.4) $\qquad L(x_0)f = (Af)(x) \Big|_{x=x_0}$.

Since A is bounded, L is defined for all $f \in L^2(a,b)$; it is linear and because of (12.1), bounded. Therefore, according to the Lemma of F. Riesz, there exists a $\alpha_{x_0} \in L^2(a,b)$, such that

(12.5) $\qquad L(x_0)f = [\alpha_{x_0}, f]$.

Since (12.5) holds for all $a \leq x_0 \leq b$, we can write

(12.6) $\qquad L(x)f = (Af)(x) = [\alpha_x, \bar{f}] = \int_a^b \alpha(x,y)f(y)dy$

where $\alpha(x,y) = \alpha_x(y) \in L^2(a,b)$; therefore (6.11) is satisfied. For normal operators, Lemma 12.1 implies, according to Theorem 8.1, that A contains zero as a Weyl limit point in the spectrum. This completes the proof.

13. CONCLUDING REMAKRS. OPEN QUESTIONS

Let us conclude with a few topics which could have been included in the material of the seminar, but for the lack of time, and with some open questions.

Linear operators on L^2, which are not necessarily Carleman operators, were touched only once, in Sec. 5, as the superposition of a continuous group of substitution operators, $\int_{-\infty}^{\infty} d\nu \rho(\nu) H^\nu$. Further investigation would reveal that — while operators of this superposition form can be written as ordinary integral operators, and even as Carleman operators if ρ and H satisfy certain conditions — on the other hand Carleman operators cannot, in general, be written as superpositions of operators of type H.

We had no time to discuss another class closely related to integral operators — smoothing and averaging operators. Thus, we have to satisfy ourselves with the few hints we have made, in particular, that the inverses of these smoothing operators — if they exist — satisfy multiplication theorems of the kind we have studied in detail.

Discussion of algebras of functions of several variables had to be omitted as well. While in one variable the linear substitution $Hf(x) = f(\alpha x)$ is relatively trivial, in several variables we have $Hf(x) = f(Mx)$, where M is a n × n matrix, and even this elementary part of the theory becomes more interesting. In particular, we could have seen, in detail, characterization of the regular analytic functions as invariants of a certain simple substitution operator. (Cf. Remark No. 2 in Sec. 2.)

Finally, we had to omit convolution algebras, where convolution takes the place of point-wise multiplication. A study of H-type Bourlet operators, i.e., especially, linear endomorphisms – which leads to substitution operators for point-wise multiplication – leads to quite another type of operation on Faltung algebras.

On this multiplication class, the non-commutative operation is

$$(13.1) \qquad \omega_1(x,y) * \omega_2(x,y) = \int_a^b \omega_1(x,\xi) \omega_2(\xi,y) d\xi,$$

a generalization of convolution; and operators of the form (11.7) are endomorphisms.

We now list several open problems; some, but certainly not all of these, may be quite easy to answer.

I. a) What is the answer to the three principal questions of Sec. 1 (cf. Sec. 2) in the case of the algebra of entire functions of one complex variable?

b) And in the case of the algebra $C(0,1)$ of functions continuous on the closed unit interval?

c) Is it possible to characterize a function algebra according to the answers to the principal questions?

II. a) (Sec. 3). Can one embed any system of generalized Chebyshev polynomials in a continuous semi-group of type (3.29)?

b) Do all elements of a system of generalized Chebyshev polynomials have a common fixed point? The answer to this question is affirmative, if the answer to IIb is affirmative.)*

c) Give necessary and sufficient conditions on the function R such that the recursion (3.24) should define a system of polynomials. (The obvious sufficient condition, that R be a polynomial in x and y, is too narrow and leads, as seen, to essentially one system, (3.32); even the Chebyshev polynomials are derived from the irrational addition theorem of the cosine. Ritt's Lemma 3.2 might lead to an answer.)

III. Establish the heuristic relationship (4.24), rigorously one some suitable function algebra.

IV. (Sec. 8). Is the product of two Carleman operators always a Carleman operator?

V. (Sec. 9). Do unbounded, strong Carleman operators exist? If yes, what can be said about the kernels? We know, of course, that such operators cannot be Hilbert-Schmidt operators; their spectrum must have Weyl limit points at zero and at infinity and nowhere else.

VI. Let $N = \int dy\, \nu(x,y)$ be a normal Carleman operator, such that $N' = \int dy\, \overline{\nu(y,x)}$ is not a Carleman operator. We know from Lemma 8.7: $N^* = \int dy \nu^*(x,y)$ is a Carleman operator. On the other hand,

---

*A more general question which has been unsolved for many years is this: Do any two commuting continuous mappings of (0,1) into itself have a common fixed point? (See, e.g. [20] and [35].) Recent, still unpublished, results found by J.P. Huneke at Wesleyan University and, independently, by W.M. Boyce a Tulane University indicate the answer to this question is negative. See [36], [37].

$$\nu^*(x,y) \neq \overline{\nu(y,x)},$$

since the r.h.s. does not satisfy the Carleman condition (6.11). What can be said about the relationship between $\nu(x,y)$ and $\nu^*(x,y)$? Construct at least one example on which the situation can be studied; in particular, do normal Carleman operators exist for which the kernel $\nu(x,y)$ is not an L^2-function as a function of x?

VII. Consider the Carleman operator A and the class E of all Carleman operators unitarily equivalent to A. This can be done by first considering UAU*, where U takes all elements of the unitary group, and then removing those elements which are not Carleman operators. (If A is a Hilbert-Schmidt operator, there will be no elements to remove.) Now consider the set G of all unitary operators V (it can be seen to be a subgroup of the unitary group) which carry E into itself, i.e.,

$$VAV^* \in E, \text{ if } A \in E.$$

Characterize G in terms of A, resp. the kernel of A. (A characterization in terms of the spectrum of A would be most appropriate, since the choice of A, and therefore the kernel of A, from among the elements of the unitary equivalence class, is accidental. This question was formulated by D. Speiser in connection with [26].)

VIII. Consider the Carleman operator $A = \int dy\, \alpha(x,y)$ and the semi-group S of all unitary operators U such that UAU* is again a Carleman operator. (As opposed to VII, now the choice of A is important.) Characterize S in terms of A. Moreover, is S identical with the semi-group S' of unitary operators such that $\int |(CU^*E)^2 \alpha(x,y)|^2 dy < \infty$ for almost all x? (Cf. (11.7).)

LIST OF REFERENCES

[1] Bourlet, C., C.R. Acad. Sci., Paris, 124 (1897), 1431-1433.

[2] Pincherle, S. and Amaldi, U., Le operazioni distributive, Bologna, Zanichelli, 1901.

[3] Kordylewski, J. and Kuczma, M., Ann. Polon. Math., 9 (1960-1961), 119-136.

[4] Rota, G.-C., Reynolds Operators. Proceedings of Symposia in Applied Mathematics, Vol. XVI, Stochastic Processes in Mathematical Physics and Engineering, 1964.

[5] Schröder, E., Math. Ann., 3 (1871), 296-322.

[6] Sansone, G., and Gerretsen, J., Lectures on the Theory of Functions of a Complex Variable, Vol. I., Holomorphic Functions, Groningen, 1960.

[7] Ikebe, T., Archive Rat. Mech. Anal., 5 (1960), No. 1, 1-34.

[8] Hille, E. and Phillips, R.S., Functional Analysis and Semi-Groups, AMS Colloquium Publications, Revised Edition, 1957.

[9] Rickart, Ch. E., General Theory of Banach Algebras, Van Nostrand, 1960.

[10] Koopman, B.O., Proc. Nat. Acad. Sci., USA, 17 (1931), 315-318.

[11] Bourlet, C., Ann. Sci. Ec. Norm. Sup., (3) 14 (1897), 133-190.

[12] Meisters, G.H., Local linear dependence and the vanishing of the Wronskian, RIAS Technical Report, No. 60 - 22, 1960.

[13] McKiernan, M.A., Publ. Math. Decrecen, 10 (1963), 30-39.

[14] Szekeres, G., Acta Math., 100 (1959), 203-258.

[15] Kuczma, M., On the Schröder equation, Rozprawy Matematyczne XXXIV, Warszawa, 1963.

[16] Aczel, J., Vorlesungen über Funktionalgleichungen und ihre Anwendungen, Birkhäuser, 1961. [English translation, with much new material, 1966, Academic Press, New York.]

[17] Loomis, L.H., An Introduction to Abstract Harmonic Analysis, van Nostrand, 1953.

[18] Ritt, J.F., Trans. Amer. Math. Soc., 23 (1922), 51-66.

[19] Targonski, Gy., Transac. New York Acad. Sci., (2) 27 (1965), 600-605.

[20] Shields, A.L., Proc. Amer. Math. Soc., 15 (1964), 703-706.

[21] Epstein, B., Orthogonal Families of Analytic Function, Mac Millan, 1965.

[22] Carleman, T., Sur les équations singulières à noyaux réelles et symétriques, Uppsala, 1923.

[23] Weyl, H., Rend. del Circ. Math. di Palermo, 27 (1909), 373-392.

[24] Neumann, John von, Charakterisierung des Spektrums eines Integraloperators, Collected Works, Vol. IV, 38-55.

[25]   Jauch, J.M., Helv. Phys. Acta, 31 (1958), 127; 31 (1958), 661.

[26]   Misra, B., Speiser, D., and Targonski, Gy., Integral Operators in the Theory of Scattering, Helv. Phys. Acta, 36 (1963), 963-980.

[27]   Schreiber, M., Acta Sci. Math. (Szeged), 24 (1963), 82-87.

[28]   Miller, J.B., Averaging and Reynolds Operators on Banach Algebras I. Representation by derivations and antiderivations. (Research Report).

[29]   Stone, M.H., Linear Transformations in Hilbert Space., Am. Math. Soc. Coll. Publ., 1932.

[30]   Dixmier, J., Algèbres de von Neumann, Gauthier-Villars, Paris, 1957.

[31]   Zaanen, A.C., Linear Analysis, North Holland, 1956.

[32]   Eidelheit, M., Studia Math., 9 (1940), 97-105.

[33]   Targonski, Gy., On a Theory of Linear Functional Equations. (To appear in Proc. 14. Scandinavian Math. Congress, Copenhagen, 1964).

[34]   Pincherle, S., Funktionaloperatoren und -gleichungen, Enzykl. der math. Wiss. II.A., 11, Vol. II. 1. (1906), 761-817. [Enlarged French version of the same article in Encycl. des sciences math. pures et appl. II. 1, Vol. II. 26 (1912)].

[35]   Baxter, Glen, Proc. Amer. Mat. Soc., 15 (1964), 851-855.

[36]   Huneke, J.P., Notices AMS, February 1967, p. 284.

[37]   Boyce, W.M., Notices AMS, February 1967, p. 280.

Lecture Notes in Mathematics

Bisher erschienen/Already published

Vol. 1: J. Wermer, Seminar über Funktionen-Algebren.
IV, 30 Seiten. 1964. DM 3,80

Vol. 2: A. Borel, Cohomologie des espaces localement
compacts d'après J. Leray.
IV, 93 pages. 1964. DM 9,-

Vol. 3: J. F. Adams, Stable Homotopy Theory.
2nd. revised edition. IV, 78 pages. 1966. DM 7,80

Vol. 4: M. Arkowitz and C. R. Curjel, Groups of Homotopy
Classes. 2nd. revised edition. IV, 36 pages. 1967.
DM 4,80

Vol. 5: J.-P. Serre, Cohomologie Galoisienne.
Troisième édition. VIII, 214 pages. 1965. DM 18,-

Vol. 6: H. Hermes, Eine Termlogik mit Auswahloperator.
IV, 42 Seiten. 1965. DM 5,80

Vol. 7: Ph. Tondeur, Introduction to Lie Groups
and Transformation Groups.
VIII, 176 pages. 1965. DM 13,50

Vol. 8: G. Fichera, Linear Elliptic Differential
Systems and Eigenvalue Problems.
IV, 176 pages. 1965. DM 13,50

Vol. 9: P. L. Ivănescu, Pseudo-Boolean Programming and
Applications. IV, 50 pages. 1965. DM 4,80

Vol. 10: H. Lüneburg, Die Suzukigruppen und ihre
Geometrien. VI, 111 Seiten. 1965. DM 8,-

Vol. 11: J.-P. Serre, Algèbre Locale. Multiplicités.
Rédigé par P. Gabriel. Seconde édition.
VIII, 192 pages. 1965. DM 12,-

Vol. 12: A. Dold, Halbexakte Homotopiefunktoren.
II, 157 Seiten. 1966. DM 12,-

Vol. 13: E. Thomas, Seminar on Fiber Spaces.
IV, 45 pages. 1966. DM 4,80

Vol. 14: H. Werner, Vorlesung über Approximations-
theorie. IV, 184 Seiten und 12 Seiten Anhang. 1966. DM 14,-

Vol. 15: F. Oort, Commutative Group Schemes.
VI, 133 pages. 1966. DM 9,80

Vol. 16: J. Pfanzagl and W. Pierlo, Compact Systems
of Sets. IV, 48 pages. 1966. DM 5,80

Vol. 17: C. Müller, Spherical Harmonics.
IV, 46 pages. 1966. DM 5,-

Vol. 18: H.-B. Brinkmann und D. Puppe, Kategorien
und Funktoren.
XII, 107 Seiten. 1966. DM 8,-

Vol. 19: G. Stolzenberg, Volumes, Limits and Extensions
of Analytic Varieties. IV, 45 pages. 1966. DM 5,40

Vol. 20: R. Hartshorne, Residues and Duality.
VIII, 423 pages. 1966. DM 20,-

Vol. 21: Seminar on Complex Multiplication. By A. Borel,
S. Chowla, C. S. Herz, K. Iwasawa, J.-P. Serre.
IV, 102 pages. 1966. DM 8,-

Vol. 22: H. Bauer, Harmonische Räume und ihre Potential-
theorie. IV, 175 Seiten. 1966. DM 14,-

Vol. 23: P. L. Ivănescu and S. Rudeanu, Pseudo-Boolean
Methods for Bivalent Programming.
120 pages. 1966. DM 10,-

Vol. 24: J. Lambek, Completions of Categories. IV, 69 pages.
1966. DM 6,80

Vol. 25: R. Narasimhan, Introduction to the Theory of
Analytic Spaces. IV, 143 pages. 1966. DM 10,-

Vol. 26: P.-A. Meyer, Processus de Markov. IV, 190
pages. 1967. DM 15,-

Vol. 27: H. P. Künzi und S.T. Tan, Lineare Optimierung
großer Systeme. VI, 121 Seiten. 1966. DM 12,-

Vol. 28: P. E. Conner and E. E. Floyd, The Relation of
Cobordism to K-Theories. VIII, 112 pages.
1966. DM 9,80

Vol. 29: K. Chandrasekharan, Einführung in die
Analytische Zahlentheorie. VI, 199 Seiten.
1966. DM 16,80

Vol. 30: A. Frölicher and W. Bucher, Calculus in
Vector Spaces without Norm. X, 146 pages. 1966.
DM 12,-

Vol. 31: Symposium on Probability Methods in Analysis.
Chairman: D.A.Kappos. IV, 329 pages. 1967. DM 20,-

Vol. 32: M. André, Méthode Simpliciale en Algèbre
Homologique et Algèbre Commutative. IV, 122 pages. 1967.
DM 12,-

MIX
Papier aus verantwortungsvollen Quellen
Paper from responsible sources
FSC® C105338

If you have any concerns about our products,
you can contact us on
ProductSafety@springernature.com

In case Publisher is established outside the EU,
the EU authorized representative is:
**Springer Nature Customer Service Center GmbH
Europaplatz 3, 69115 Heidelberg, Germany**

Printed by Libri Plureos GmbH
in Hamburg, Germany